T0211126

Semiconductor Packaging

Materials

Interaction

and

Reliability

Semiconductor Packaging

Materials

Interaction

and

Reliability

Andrea Chen
Randy Hsiao-Yu Lo

CRC Press
Taylor & Francis Group
Boca Raton London New York

CRC Press is an imprint of the
Taylor & Francis Group, an **informa** business

CRC Press
Taylor & Francis Group
6000 Broken Sound Parkway NW, Suite 300
Boca Raton, FL 33487-2742

First issued in paperback 2017

© 2012 by Taylor & Francis Group, LLC
CRC Press is an imprint of Taylor & Francis Group, an Informa business

No claim to original U.S. Government works

Version Date: 20110819

ISBN 13: 978-1-4398-6205-6 (hbk)
ISBN 13: 978-1-138-07540-5 (pbk)

Visit the Taylor & Francis Web site at
http://www.taylorandfrancis.com

and the CRC Press Web site at
http://www.crcpress.com

Contents

Section III: Materials used in semiconductor packaging

Preface

Semiconductor packaging assembly and testing is an important manufacturing step necessary to create electronic products. The subject is not understood in much depth, as compared to a subject like circuit design. It is generally believed that this is due to the nature of the topic, as back-end processing of semiconductors is a multidisciplinary area, encompassing materials science, mechanical design, electrical layout and modeling, and many other engineering specialties. This book specifically addresses that shortcoming, especially in the area of materials used for packaging semiconductors and their interactions.

Simply put, semiconductor packages are not monolithic entities but are made up of many different components interlocked with the common goal of protecting the integrated circuit while allowing it to communicate with the outside world—the lead frame or substrate on which the chip sits upon, the die attach adhesive to glue the chip down, electrical connections made via metallic bonding wires or bond pad bumps for flip-chip attachment, and finally an epoxy molding compound to encapsulate everything for protection. And, those components are made up of very different materials: polymers, composites, metals, and various combinations of those categories. Understanding how the various materials behave and interact as they form the protective barrier for the delicate chip is key for making the package reliable and robust.

To gain these insights, a basic knowledge of material properties is necessary, along with determining which behaviors are important to package performance. That takes understanding how a given property is measured and why it is significant. For example, the measurement of viscosity versus time—a viscosity curve—in a molding compound or the length of the heat-affected zone in metallic wire used in thermosonic bonding eventually translates into a certain level of performance for manufacturing or reliability purposes for a given package.

From there, the next step is looking at how these properties of various packaging materials interact with one another and how to maximize their performance in regard to package integrity and reliability. Again, as an example, the length of the heat-affected zone in a bonding wire will help determine how high or how low the final loop height will be. In another example, a viscosity curve for a molding compound often acts as an indicator of its utility in manufacturing, by its ability to efficiently and completely fill cavities in a mold chase.

This book is focused on providing a fundamental understanding of the underlying physical properties that make up the materials used in a semiconductor package. By tying together the disparate elements that are essential to a semiconductor package, this volume hopes to convey the knowledge of how all the parts fit and work together to provide durable protection to the integrated circuit chip within as well as a means for the chip to communicate with the outside world.

Authors

Andrea Chen received a B.S. in Materials Science and Engineering from University of California, Berkeley and a M.S. in Materials Engineering from Rensselaer Polytechnic Institute, New York. She started her career at National Semiconductor Corporation in the Package Technology group, working on various semiconductor packaging materials and reliability issues. Subsequently, she went to ChipPAC, Inc., with the Technology Development group, involved in low-cost flip-chip technology development. Currently, Chen works at Siliconware USA, Inc. (SPIL) as a technical marketing manager. To date, she has coauthored more than 30 papers and presentations.

Randy Hsiao-Yu Lo received his B.S. from National Taiwan University, Taiwan in 1979; M.S from Worcester Polytechnic Institute, Massachusetts in 1984; and Ph.D. from Purdue University, Indiana in 1990—all in the field of chemical engineering. He was at National Semiconductor Corporation from 1984 onward, eventually becoming senior engineering manager in the Package Technology group. Subsequently, he headed the Electronic Packaging Development group at ERSO/ITRI-Taiwan from 1997 to 1998. Then, he led the Siliconware Precision Industries Limited (SPIL) Research and Development group until 2000. Later that same year, he was appointed executive vice president and head of North America Sales and Marketing for Siliconware USA, Inc. In 2001, Lo was appointed president for Siliconware USA, Inc.—the position he currently holds. To date, he has coauthored more than 20 papers and presentations and is listed as coinventor on over 30 U.S. patents.

Partial list of abbreviations, acronyms, and symbols

ACLV—autoclave
AES—Auger electron spectroscopy
AlN—aluminum nitride
Al_2O_3—alumina, aluminum oxide
ASIC—application-specific integrated circuit
CMOS—complementary metal oxide semiconductor
CCD—charge coupled device
CSP—chip scale package
CTE—coefficient of thermal expansion
DIP—dual in-line (through-hole leads) package
DRAM—dynamic random access memory
DSC—differential scanning calorimetry
DSP—digital signal processor
EFO—electronic flame off
EIA—Electronics Industries Alliance
EDS or EDX—energy-dispersive X-ray spectroscopy
ENIG—"Electroless Nickel Immersion Gold"
ESD—electrostatic discharge
FAB—free air ball
FPGA—field programmable gate arrays
FTIR—Fourier transform infrared spectroscopy
HAST—highly accelerated stress test
HAZ—heat-affected zone
HDI—high-density interconnect
HTSL—high-temperature storage life
Hz—hertz, as in one cycle per second (1 Hz = 1 cycle per second)
IC—integrated circuit
IQA—incoming quality assurance
I/O—input-output
IR—infrared
κ—dielectric constant
LED—light-emitting diode

LTCC—low-temperature co-fired ceramic
MCM—multichip module
MCP—multichip package
MEMS—Micro Electro-Mechanical System
OSP—organic solderability preservative
PCB—printed circuit board
PGA—(through-hole) pin grid array packages
PLCC—plastic leaded chip carriers, with surface-mount J-leads
PoP—package-on-package
ppm—parts per million
PQFP—plastic quad flat pack
PTH—plated through-hole
RAM—random access memory
RGB—red-green-blue (LED lighting combination)
SAM—scanning acoustic microscopy
SEM—scanning electron microscopy
SIMS—secondary ion mass spectrometry
SiO$_2$—silica, silicon dioxide
SO or SOP—small outline package, with gull-wing surface-mount leads
SOJ—small outline package, but with surface-mount J-leads
SMT—surface-mount technology
TAB—tape automated bonding
TEM—transmission electron microscopy
T_g—glass transition temperature
THB—temperature-humidity-bias reliability test
TMCL—thermal or temperature cycling reliability test
TOP—transistor outline package
TQFP—thin quad flat pack
TSOP—thin small outline package
YAG—ytterium-aluminum-garnet (phosphor used in LED production)

Bibliography

C.A. Harper, *Electronic Packaging and Interconnection Handbook*, McGraw-Hill Professional, New York, 1991.

L.T. Nguyen, R.H.Y. Lo, A.S. Chen, and J.G. Belani, "Molding Compound Trends in a Denser Packaging World II: Qualification Tests and Reliability Concerns," *IEEE Trans. on Reliability*, vol. 42, no. 4, 518–535, December 1993.

S.L. Oon, "The Latest LED Technology Improvement in Thermal Characteristics and Reliability—Avago Technologies' Moonstone 3-in-1 RGB High Power LED," *Avago Technologies White Paper*, AV02-1752EN, March 17, 2010.

M. Wright, "Intematix Launches New Red and Green LED Phosphors," *LEDs Magazine*, November 11, 2010.

section one

Semiconductor packages

chapter one

History and background

1.1 Objectives

- Discover semiconductor packaging.
- Provide brief background and history of semiconductor packaging technology.
- Learn basic process steps involved in plastic semiconductor package assembly.

1.2 Introduction

Semiconductor packaging is a middle link in electronics systems manufacturing, starting from wafer fabrication of multiple integrated circuits and proceeding all the way to final enclosure for the finished product. It must meet the demands of the steps prior to it, at the front end of production, and those steps that follow, through mounting on a printed circuit board and final systems integration. These demands tend to be contradictory, while at the same time, requirements for increased and better performance from the package are always increasing.

Simply put, a semiconductor package is a semiconductor chip enclosed or encapsulated to assure environmental protection, and it provides for a reliable means of interconnection to the next level of integration. The package is dubbed the *first level* of packaging, with the circuit board being the *second level* and the final enclosure the *third level*.

Specifically, a semiconductor package should protect the chip from mechanical stresses (vibration, falling from a height), environmental stresses (such as humidity and contaminants), and electrostatic discharge (also known as ESD) during handling and mounting onto a printed circuit board and beyond. In addition, the package is the mechanical interface for electrical testing, burn-in, and the next level of interconnection. Last, the package must also meet the chip's various performance requirements, encompassing the physical, mechanical, electrical, and thermal. Finally, the package must meet specifications for quality and reliability as well as be a cost-effective solution toward the final product. In all, semiconductor packages are an important part of any electronics system, though it is often neglected or treated as an afterthought—until there is a problem.

Therefore, this chapter addresses background information needed to understand the use and importance of the various materials and components used in semiconductor packaging.

1.3 Brief history

Here is a brief description of the progression of semiconductor packages, from metal cans and ceramic packages in the early days to today's packages made up of lightweight organic materials. In the future, even more exotic materials may be commonplace in complex package structures.

1.3.1 Hermetic packaging

As already mentioned, in the early days of the semiconductor industry, the majority, if not all, of semiconductor packages were ceramic based or metal cans. Given that the earliest adopters of semiconductors were the military and aerospace industries, hermetic packages offered the highest levels of reliability under any possible adverse operating conditions. By design, a hermetic seal prevents any contaminants, whether gases, liquids, or particulates, from reaching the sensitive and relatively delicate semiconductor chip surface within the package cavity. More details on hermetic packages are given in Chapter 2.

However, their robustness also had drawbacks. Hermetic materials tend to be costly and hard to manufacture and process, due to their hardness and brittle natures. The packages—the ceramic ones, especially— could be heavy and large, which meant the printed circuit board and overall enclosure had to also be large and heavy to support the weight. Finally, hermetic packages tend not to lend themselves to miniaturization. The semiconductor industry eventually turned to using organic materials and plastics, for both cost and weight savings, starting in the 1970s.

Though plastic package unit volumes now far surpass those for hermetic ones, they remain in use for applications that have demanding performance and environmental needs. Two applications where hermetic packaging still finds demand are light-emitting diode (LED) and Micro Electro-Mechanical System (MEMS) packaging, which is discussed in Chapter 4.

Many of the well-known and commonly used semiconductor packages that will be described shortly have both ceramic originators and organic successors. They include dual in-line, pin grid arrays, leaded chip carriers, and they all have ceramic and molding compound versions.

1.3.2 Plastic packaging

The plastic version of the dual in-line packaging (DIP) family was introduced in the early 1970s and then proceeded to dominate the market for plastic packages until the late 1980s, when surface-mount technology (SMT) arrived on the scene, with quad flat packs, small outline packages, and plastic leaded chip carriers.

To summarize the progress of package technology, through-hole packages like DIPs were the first plastic semiconductor packages to enter widespread use. "Through-hole" refers to the fact the package leads or pins went through holes in the printed circuit board to make their physical and electrical connections. Packages and their leads were both rather large in size. The next shift in technology came with the development of surface-mount technology, where the package leads were connected to lands on the circuit board, which allowed for both package body size and the leads to shrink in size. This is discussed in more detail in Chapter 3. More recently, the increasing pin counts forced the development of area array packages, like the ball grid array (BGA). The trend to ever-smaller, lighter, and thinner consumer electronic products was only achievable by further miniaturization, which drove the concept of chip scale packaging (CSP). The next step to reduce packaging cost and size was the approach of finishing the chip package directly on the wafer; thus, wafer-level packaging (WLP) was created. Going forward, this book will focus on plastic semiconductor packages grouped by interconnect method, whether by wire bonding or flip-chip attachment, and finally under wafer-level packaging.

Though in the past, packaging and interconnection technologies were not limiting factors in wringing maximum performance out of a device, demands from both the device and system ends has meant more and more focus is being placed on package technology as a limiting factor, and the industry is looking for ways to address these issues.

1.4   Wire bonding process flow

The fundamental process flow for semiconductor packaging using wire bonding has remained relatively unchanged for the past 40 and more years, though the equipment and materials used have undergone considerable improvement and changes. Manufacturing operations have gone from manual, labor-intensive operations to highly automated, high-volume production. Materials are of higher overall quality and chemical purity, and they are engineered for specific properties and applications. Table 1.1 shows the process steps associated with wire-bonded packages, both plastic (such as for a plastic ball grid array) and hermetic (such as for ceramic dual-inline packages) types, and Figure 1.1 illustrates the process flow.

1.5   Flip-chip process flow comparison

Table 1.2 compares the process steps changes going from a wire-bonded substrate package to one using a flip chip. In addition to starting with

Table 1.1 Comparison of Process Steps between Package Types

Plastic (Lead Frame)	Plastic (Laminate)	Hermetic
Wafer sort	Wafer sort	Wafer sort
Second optical	Second optical	Second optical
Wafer mount	Wafer mount	Wafer mount
Wafer sawing	Wafer sawing	Wafer sawing
Die attach	Die attach	Die attach
Wire bond	Wire bond	Wire bond
Third optical	Third optical	Third optical
Encapsulate (molding compound)	Encapsulate (molding compound or glob top)	Lid seal
Dejunk	Ball attach and reflow	Leakage test
Deflash	Singulate	Marking
Marking	Ball inspection	
Plating	Marking	
Trim and form		
Final inspection	Final inspection	Final inspection

Source: Adapted from National Semiconductor Corporation, *Data Sheet: Semiconductor Packaging Assembly Technology*, August 1999.

a bumped wafer, the major changes come in during the interconnection steps—die attach adhesive plus wire bond replaced pick-and-place plus reflow. Type of encapsulation is also different, with molding compound generally replaced by underfill.

Further discussion on underfill materials is given in Chapter 6, Section 6.3, and more information on wafer bumping will be presented in Chapter 7, Section 7.4.

1.6　Equipment

As an example, Figure 1.2 illustrates a transfer mold press used with molding compound to encapsulate plastic packages. Nearly all the steps shown in Table 1.2, which describe semiconductor package assembly and manufacturing, now use automated equipment for volume production—the die bonder, the wafer saw, and, of course, the wire bonder, to name a few examples.

1.7　Material interactions

The heterogeneous components that make up a semiconductor package often differ wildly in physical properties, as shown in Table 1.3. The key is to find the most reasonable material set that is also cost-effective and manufacturable in high-volume assembly. These interactions will be discussed in detail in subsequent chapters.

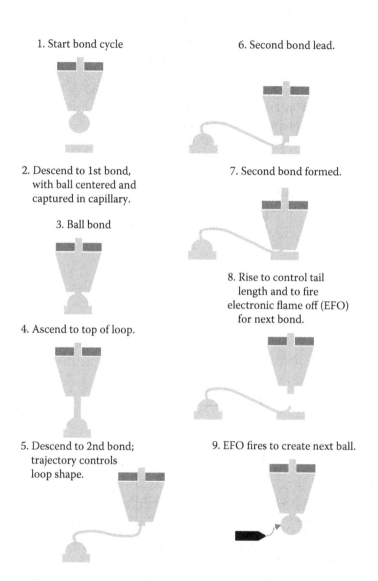

1. Start bond cycle

2. Descend to 1st bond, with ball centered and captured in capillary.

3. Ball bond

4. Ascend to top of loop.

5. Descend to 2nd bond; trajectory controls loop shape.

6. Second bond lead.

7. Second bond formed.

8. Rise to control tail length and to fire electronic flame off (EFO) for next bond.

9. EFO fires to create next ball.

Figure 1.1 Process steps in gold thermosonic wire bonding.

Table 1.2 Wire Bond versus Flip-Chip
Process Flows for a Substrate Package

Wire Bond	Flip Chip
Wafer	Wafer
Dice	Wafer bumping
Die attach	Dice
Cure	Pick and place plus flux
Wire bonding	Reflow
Encapsulate	Underfill encapsulation
Ball attach	Ball attach
Mark	Mark
System test	System test

Source: Adapted from P. Elenius and L. Levine,
"Comparing Flip-Chip and Wire-Bond
Interconnection Technologies," *Chip Scale
Review*, 81–87, July/August 2000.

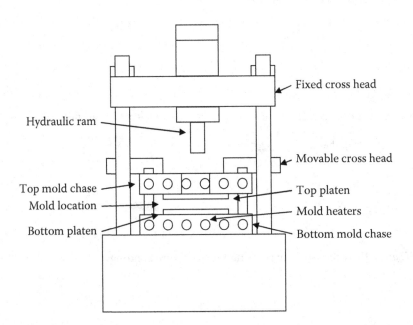

Figure 1.2 Transfer mold press.

Table 1.3 Key Properties of Semiconductor Packaging Materials

Material	CTE (ppm/°C)	Density (g/cm³)	Thermal Conductivity (W/m*K)	Electrical Resistivity (μΩ-cm)	Tensile Strength (GPa)	Melting Point (°C)
Silicon	2.8	2.4	150	N/A	N/A	1430
Molding Compound	18–65	1.9	0.67	N/A	N/A	165 (T_g)
Copper	16.5	8.96	395	1.67	0.25–0.45	1083
Alloy42	4.3	N/A	15.9	N/A	0.64	1425
Gold	N/A	19.3	293	2.2	N/A	1064
Aluminum	23.8	2.80	235	2.7	83	660
Eutectic Tin-Lead Solder	23.0	8.4	50	N/A	N/A	183
Alumina	6.9	3.6	22	N/A	N/A	2050
Aluminum nitride	4.6	3.3	170	N/A	N/A	2000

Source: Adapted from National Semiconductor Corporation, *Data Sheet: Semiconductor Packaging Assembly Technology*, August 1999.

Bibliography

M.G. Bevan and B.M. Romenesko, "Modern Electronic Packaging Technology," *Johns Hopkins APL Technical Digest*, vol. 20, no. 1, 22–33, 1999.

P. Elenius and L. Levine, "Comparing Flip-Chip and Wire-Bond Interconnection Technologies," *Chip Scale Review*, 81–87, July/August 2000.

K. Gilleo, B. Cotterman, and T. Chen, "Molded Underfill for Flip Chip in Package," *HDI Magazine*, June 2000.

C.A. Harper, *Electronic Packaging and Interconnection Handbook*, McGraw-Hill Professional, New York, Chapter 6, 1991.

National Semiconductor Corporation, *Data Sheet: Hermetic Packages*, August 1999.

National Semiconductor Corporation, *Data Sheet: Semiconductor Packaging Assembly Technology*, August 1999.

M. Osborne, "A Comprehensive Study of Fine-Pitch Bonding Reveals the Importance of Process Control," *Chip Scale Review*, March 2006.

M. Töpper, "10th Anniversay Insights—A Short History of Wafer-Level Packaging," *Advanced Packaging*, April 2002.

R.R. Tummala, "Electronic Packaging Research and Education: A Model for the 21st Century," *Johns Hopkins APL Technical Digest*, vol. 20, no. 1, 111–121, 1999.

chapter two

Package form factors and families

2.1 Objectives

- List and categorize the different types of plastic semiconductor packages currently available on the market.
- Discuss the various failure modes plastic packages are subject to, and potential remedies.

2.2 Introduction

The variety of plastic semiconductor packages available in the industry only continues to proliferate over time. Even as new types and form factors come into being, the established players do not disappear entirely, given their long, reliable history and cost-competitiveness when price and not performance is the primary factor for package selection. Table 2.1 shows the mature package families of dual in-line, small outline, and thin small outline packages still account for over 40% of semiconductor package units produced worldwide.

The increase in the number of package types matches the growth seen in the number and types of electronic products entering into common use. There are an ever-growing number of new applications for personal, healthcare, home, automotive, security, and entertainment systems. Advancement in package technology helped create the innovative solutions needed in new and future products.

2.3 Package outline standardization

The industry association JEDEC—formed in 1958 as the Joint Electron Devices Engineering Council and now officially known as *JEDEC Solid State Technology Association*—regulates the standards and drawings for package configurations, outlines. JEDEC is the standardization arm of the *Electronics Industries Alliance (EIA)* and is a member of that umbrella organization. The comprehensive guide to all registered package outlines is in Publication 95 (JEP95), located on the JEDEC website.

Table 2.1 2007 Worldwide Integrated Circuit (IC) Packaging Units by Package Family

Package Type	Share, %	
SO	24.8	
TSOP	13.4	
SOT	7.6	
DIP	5.3	
DCA	7.7	
WLP	4.2	
FBGA/DSBGA	13.7	
BGA	4.4	
PGA	0.1	
QFN	5.4	
QFP	9.2	
CC	0.8	
DFN	3.5	
Total	100	151 billion units

Notes: SO, small outline package; TSOP, thin small outline package; SOT, small outline transistor; DIP, dual in-line (through-hole leads) package; DCA, direct chip attach; WLP, wafer-level packaging; FBGA, fine pitch ball grid array/DSBGA, die-sized ball grid array; BGA, ball grid array; PGA, pin grid array packages; QFN, quad flat no lead; QFP, quad flat pack; CC, chip carrier; DFN, dual flat no lead.

Source: Adapted from Sandra Winkler, "Trends in IC Packaging and Multicomponent Packaging," *IEEE SCV Components, Packaging and Manufacturing Technology Chapter*, January 22, 2009.

2.4   Leaded package families

As noted in Chapter 1, leaded packages have been used in the industry for decades. Even though the styles of packages have proliferated, they remain in wide use, especially for low-pin count parts and when cost is a primary consideration.

2.4.1   Dual lead package family

Dual leaded packages are based on mature two-sided lead frame technology utilizing either PTH (plated through hole) or SMT (surface-mount technology). Within SMT grouping, there is *J lead*, which folds underneath the package body, and *gull wing*, where the leads fan away from the package body—configurations available for different SMT requirements. Lead counts range up to 86 pins.

There are a multitude of package types within the dual lead family, depending on lead type and shape, lead pitch, and body size and thickness. The plastic dual in-line package (or PDIP) uses PTH board mounting technology. Of the dual lead packages utilizing SMT, they are all variations of the small outline package (SOP), with the acronyms and names changing depending on whether the lead configuration is a J-lead (SOJ), or on the body profile and lead pitches.

2.5   Quad lead package family

Quad leaded packages are based on mature four-sided lead frame technology utilizing SMT (surface-mount technology). Within SMT grouping, there are *J-lead* and *gull-wing* lead configurations available for different SMT requirements. Lead counts range up to 256 pins.

The quad packages with J-leads are generally known as plastic leaded chip carrier (PLCC) and have lead counts up to 84. Those with gull-wing leads are known as quad flat pack (QFP) packages. The QFP subgroup has several variations, depending on body thickness and lead pitch, with lead counts ranging up to 256. QFP packages are also available with thermal enhancement, such as an exposed heat spreader.

Figure 2.1 shows a J-lead plastic package in a cross-sectional view, and Figure 2.2 is of a gull-wing leaded plastic package.

2.6   Substrate-based package families

At some point, an input/output (I/O) limitation was met with lead-frame-based packages. The leads could only be made so narrow before they became too fragile to handle during the assembly processes and final trim-and-form. To achieve greater I/O density to match the shrinking

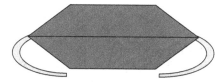

Figure 2.1 J-lead leaded plastic package (not to scale).

Figure 2.2 Gull-wing leaded plastic package (not to scale).

geometries of the chip, packages were developed to mimic printed circuit boards (PCBs), where denser electrical connections and routing could be met. Another advance was multitier wire bonding, to allow for staggered bonding lands to meet finer pitch densities. Finally, the use of organic substrates would allow flip-chip interconnect technology to become low(er) cost and more widely available for many different applications and devices.

2.6.1   Ball grid array package family

In a ball grid array (BGA) package, the chip is mounted to the top surface of a printed circuit board-type substrate instead of a metal lead frame. If the interconnection is wire bonding, the wires are connected to electrical traces on the substrate. Flip-chip interconnections may also be employed.

Compared to lead-frame packages, BGA packages offer superior electrical and thermal performance, higher interconnect density, and excellent surface-mount yields for high pin counts (usually above 256, which is generally the upper limit for QFPs).

A BGA is a package technology that employs a solder ball grid array matrix to make electrical input and output connections to a printed circuit board. BGAs offers improved electrical and thermal operation through multiple routing layers such as ground and power planes. The package family includes cavity-up and cavity-down designs utilizing advanced substrate technologies, as well as optional heat spreaders and heat sinks when even higher thermal dissipation is a necessity.

BGA packages are commonly used for high-performance applications such as microprocessors or controllers, application-specific integrated circuits (ASICs), digital signal processors (DSPs), gate arrays, and memory and computer chipsets. A thermally enhanced version is shown in

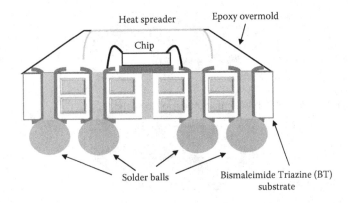

Figure 2.3 Thermally enhanced ball grid array (not to scale).

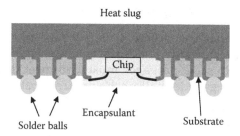

Figure 2.4 Cavity-down, thermally enhanced ball grid array (not to scale).

Figure 2.3, and Figure 2.4 shows a cavity-down version that provides both thermal and electrical improvements.

Polyimide tape can also be used as a substrate material. The tape requires attachment to a heat spreader for support as well as improved thermal performance. Drilling through the tape to allow solder ball connection to the heat spreader to act as a ground plane improves the electrical performance further, as illustrated in Figure 2.5.

2.7   Chip scale packages

A chip scale package (CSP) is defined as a package where the bare die occupies 80% or more of the package area, so the profile can be a near-chip-size package outline. Electrical performance is enhanced due to shorter interconnections. CSPs may utilize lead frames or substrates, and the substrates may be rigid or flexible. The packages may have solder balls or simply metalized lands; the lead frame version may not have external leads, as in a quad flat no-lead package (see below). Also, interconnects may be wire bonds or flip chip (see Section 2.10, Flip-chip packages).

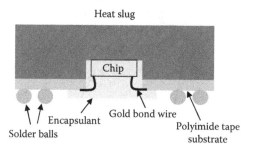

Figure 2.5 Ball grid array using polyimide tape substrate (not to scale).

2.7.1　Substrate-based chip scale packages

Essentially, substrate-based chip scale package technology is based on mature BGA technology and infrastructure, scaled down to small form factors. Chip scale packages must also meet design guidelines and rules regarding solder ball pitch and array patterns, as provided by JEDEC.

2.7.2　Quad flat no lead

A quad flat no-lead (abbreviated as QFN) package is a plastic encapsulated lead-frame-based CSP with a lead pad on the bottom of the package to provide electrical interconnection with the printed circuit board. This package offers a small form factor with 60% size reduction compared with conventional QFP packages. It provides good electrical performance due to the short electrical path in the inner leads and wires. Electrical performance may be further enhanced by using a flip chip to shorten the interconnect path even further. The package also provides excellent thermal performance by an optional exposed die pad to provide an efficient heat path soldered on the circuit board. This small and light package with improved thermal and electrical performance makes QFNs suitable for portable communication and consumer products.

As already mentioned, interconnects may be made with wire bonding or with a flip chip. An example of the wire bond type is shown in Figure 2.6, and those of the flip-chip versions are shown in Figure 2.7.

Note that *punch type* and *map type* refer to both the molding method and subsequent singulation techniques. QFN packages may be molded individually and then punched out of the lead frame like other, traditional lead-frame packages. Or, an array of individual QFN packages may be molded as one large package and singulated subsequently by sawing.

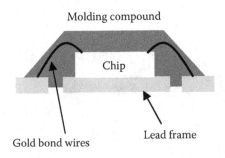

Figure 2.6 Wire-bonded quad flat no lead (not to scale).

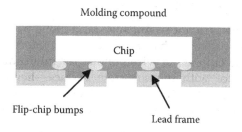

Molding compound

Chip

Flip-chip bumps

Lead frame

Figure 2.7 Flip-chip quad flat no lead (not to scale).

2.8 Stacked-die package family

The stacked-die package is a package technology that stacks multiple die vertically in the same package. For example, multiple memory devices may sit on top of one another to increase memory density, or ASICs may be combined with memory chips. Compared to single-die packages, stacked-die packages combine several different functional devices or increase memory density in the same footprints as a single-die package. Stacked-die packages may include substrate-based and lead-frame-based types.

2.9 Package-on-package and related variations

Multiple-package packages provide an alternative to multichip module solutions when known-good-die is not feasible. Multiple chips are integrated individually after undergoing functional tests within one package form factor. Rework of package on substrate is feasible to ensure the module yield. Figure 2.8 shows a variation on the multiple-packages-in-one called *package-on-package*, or PoP. In this example, the PoP vertically combines discrete memory (the top package) and logic chip (the bottom half) packages to save board space, lower overall pin count, and enhance electrical performance.

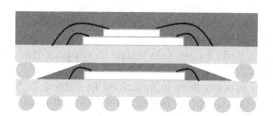

Figure 2.8 Package-on-package (not to scale).

2.10   Flip-chip packages

Flip-chip packages are not so much a stand-alone package form factor as a certain set of packages that share a common interconnect characteristic. Namely, the chip is not wire bonded for interconnect but instead flipped face-to-face with the substrate surface—hence the name *flip chip*—and the interconnection between the die and substrate is made through an array of bumps that are placed on the bonding pads of the die surface.

Flip-chip packages provide a solution for low to high pin count, high electrical performance demands from high-end memory, ASICs, and microprocessor applications where high frequency and high speed are required. Going from wire bonding to a flip-chip configuration makes it possible to jump into a higher pin count, and high electrical performance applications.

Flip-chip interconnections allow better electrical performance through lower inductances due to the shorter electrical path between the chip and the substrate. The array of bumps under the chip also allows the die to shrink in size, which can reduce wafer cost. The flip-chip structure also allows you to make power and ground connections to internal points on a die, resulting in better chip performance. To increase thermal performance, optional heat spreaders can be attached on the backside of the flipped die.

Types of packages employing a flip chip can be as small as 3 mm on a side CSPs up to mammoth 45 mm on a side BGAs. What is common among all of them is the need for enhanced electrical performance by shrinking the bond pad-to-lead distance. This can be seen in Figure 2.9, which shows the improvement in return loss in frequencies below 9 GHz between a wire-bonded quad flat no lead and a flip-chip version. Figure 2.10 shows the difference in insertion loss, which shows the flip-chip version has a bigger bandwidth range given that the wire bonded package shows 1 dB insertion loss at 10 GHz while the flip-chip package exhibits that behavior at 11 GHz.

2.11   Wafer-level chip scale packages

The concept of wafer-level packages emerged in the mid-1990s. Figure 2.11 shows a cross section from a wafer-level chip scale package. Note that redistribution is used to reroute connections from the peripheral bond pads to an area array to support the external solder ball connections.

A simplified process flow is described in Table 2.2. A finished wafer would undergo subsequent additional processing to make each individual chip a package: bond pad redistribution, additional layers of passivation, under bump metals, and finally the solder ball for external interconnection.

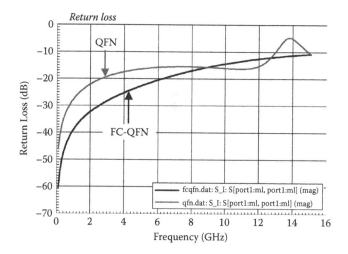

Figure 2.9 Return loss graphs comparing QFN and FC-QFN. (Reprinted with permission from Kevin Chang, Jen-Yuan Lai, Hanping Pu, Yu-po Wang, C.S Hsiao, Andrea Chen, and Randy H.Y. Lo, "Flip Chip Quad Flat No-Lead (FC-QFN)," *IWLPC 2005*, November 1, 2005.)

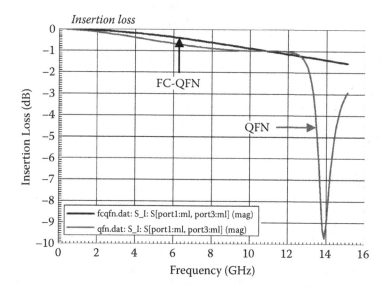

Figure 2.10 Insertion loss graphs comparing QFN and FC-QFN. (Reprinted with permission from Kevin Chang, Jen-Yuan Lai, Hanping Pu, Yu-po Wang, C.S Hsiao, Andrea Chen, and Randy H.Y. Lo, "Flip Chip Quad Flat No-Lead (FC-QFN)," *IWLPC 2005*, November 1, 2005.)

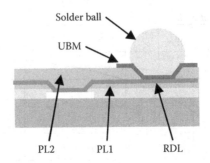

Figure 2.11 Cross section of a wafer-level chip scale package (not to scale).

Table 2.2 Process Flow for Wafer-Level Chip Scale Packages

Steps	Description
1	Finished wafer arrives from wafer fabrication
2	Passivation layer 1
3	Redistribution of bond pads
4	Passivation layer 2
5	Under bump metallurgy
6	Solder bumping
7	Solder bump reflow

It is obvious that wafer-level packaging looks similar to a flip chip, and there is often confusion on whether a bare bumped chip is a wafer-level package or just a bare flip chip. Conventions seem to dictate that a wafer-level chip scale package is one that has spherical bumps on a grid pattern with a fixed, predetermined pitch, whereas as a bare flip chip would not adhere to these rules.

Bibliography

M.G. Bevan and B.M. Romenesko, "Modern Electronic Packaging Technology," *Johns Hopkins APL Technical Digest*, vol. 20, no. 1, 22–33, 1999.

K. Chang, J.-Y. Lai, H. Pu, Y.-p. Wang, C.S. Hsiao, A. Chen, and R.H.Y. Lo, "Flip Chip Quad Flat No-Lead (FC-QFN)," *IWLPC 2005*, November 1, 2005.

A. Chen, E. Feng, R. Lo, C.-C. Wu, T.D. Her, and C.Y. Lin, "The Future in 3-D Chip Scale Packaging," *SEMICON China 2002*, March 26–27, 2002.

A. Chen, E. Feng, R. Lo, C.-C. Wu, and T.D. Her, "Recent Innovations in Stacked Die Packages," *KGD Workshop 2001*, September 12, 2001.

Freescale Semiconductor, *Application Note: Quad Flat Pack No-Lead (QFN), Micro Dual Flat Pack No-Lead (uDFN)*, AN1902, Rev. 4.0, September 2008.

C.A. Harper, *Electronic Packaging and Interconnection Handbook*, McGraw-Hill Professional, New York, Chapters 6 and 9, 1991.

M. Iyer, "Emerging Trends in Advanced Packaging," *Semiconductor International*, June 1, 2009.

JEDEC, *JEP95: Publication 95*, www.jedec.org/download/pub95/default.cfm

JEDEC, *JEP95: Publication 95, Design Guide 4.18: Wafer Level Ball Grid Arrays (WLBGA)*, Issue A, September 2004.

JEDEC, *JEP95: Publication 95, Design Guide 4.6: Fine-Pitch, Rectangular Ball Grid Array Package (FRBGA)*, Issued, April 2005.

JEDEC, *JEP156: Chip-Package Interaction—Understanding, Identification and Evaluation*, March 2009.

R.H.Y. Lo and C.-C. Wu, U.S. Patent No. 6,507,120: Flip Chip Type Quad Flat Non-Leaded Package, January 14, 2003.

R. Mahajan, K. Brown, and V. Atluri, "The Evolution of Microprocessor Packaging," *Intel Technology Journal*, Q3, 1–10, 2000.

Maxim Integrated Products, *Application Note 4002: Understanding Flip-Chip and Chip-Scale Package Technologies and Their Applications*, April 18, 2007.

National Semiconductor, *Application Note 1126: BGA (Ball Grid Array)*, August 2003.

P.J.C. Normington, "Patent Review: Patent Illustrates a New Use for Old Technology—Tape Automated Bonding," *Chip Scale Review*, July 2003.

Siliconware Precision Industries Ltd., www.spil.com.tw/

M. Töpper, "10th Anniversay Insights—A Short History of Wafer-Level Packaging," *Advanced Packaging*, April 2002.

S. Winkler, "Trends in IC Packaging and Multicomponent Packaging," *IEEE SCV Components, Packaging and Manufacturing Technology Chapter*, January 22, 2009.

chapter three

Surface-mount technology

3.1 Objectives

- Provide a brief overview of semiconductor packages used in surface-mount technology of printed circuit boards.
- Discuss how issues particular to surface-mount technology affect semiconductor packages.
- Touch upon future developments in the area of surface-mount technology.

3.2 Introduction

Surface-mount technology (SMT) refers to how various semiconductors, passives, and other components are attached to a printed circuit board (PCB), to both the top and bottom surfaces. In the beginning of the electronics industry, both packages and PCBs were relatively large in scale. Through-hole packages prevalent in the early days were called that because their leads fit into the holes drilled through the PCBs to connect power, ground, and circuitry connections.

In time, the demands of continual miniaturization of electronic products meant the boards inside them had to shrink, along with the chips and components. To save space and increase device density, the much-smaller lead area of surface-mount packages supplanted that of through-hole packages. For example, the lead pitch of a small outline package or a quad flat pack is half that or more compared to a plastic dual in-line (through-hole) package, at 1.27 mm or 1.00 mm versus 2.54 mm for the older technology. And not just for semiconductors but also for discrete devices, which lost their axial wire leads jutting out of cylindrical bodies to be replaced by much-smaller rectangular bricks with metalized end caps. Table 3.1 presents a comparison of package types and dimensions.

However, the transition from through-hole mounting technology to surface mount proved to have many challenges, not the least of which was the stress imposed on the semiconductor packages, to the point their reliability was reduced or even damaged. This chapter discusses those issues and subsequent solutions developed.

Table 3.1 Comparison of Through-Hole and Surface-Mount Package Types

Package Type	Pin Pitch	Pin Length	Body Thickness	Body Width
Plastic dual in-line	2.54 mm	3.17 mm	4.06 mm	7.62 to 22.9 mm
Small outline	1.27 mm	0.76 mm[a]	2.49 mm	3.94 to 7.62 mm
Plastic quad flat pack	1.00 to 1.27 mm	0.76 mm*	2.41 mm and below	Variable

Source: Adapted from Richard D. Skinner, Ed., *Basic Integrated Circuit Technology Reference Manual,* Integrated Circuit Engineering Corporation, Section 3: Packaging, 1993, figure 3-10.

[a] Solderable length.

3.3 Background

One major stumbling block was discovered when PCB mounting went from through-hole to surface mount, in the solder reflow step. In the days of through-hole packages and mounting, the majority of components were attached to one side of the board and the PCBs passed through a wave soldering machine, where molten eutectic tin-lead (Sn-Pb) solder was pushed along the boards backside, to wick up the through-holes to weld the package leads to the board, as illustrated in Figure 3.1. Generally, only passive components might be mounted on the bottom side and directly exposed to the solder wave, as the general wisdom advised not exposing active devices to molten solder. Having passive components only on the bottom side and through-hole active components is now generally known in the industry as a Type III SMT assembly, as shown in Figure 3.2. And while the molten metal was at least 183°C—the melting point for eutectic Sn-Pb solder—semiconductor packages unlikely saw peak temperatures very close to that level, insulated as they were by the thickness of the PCB. So, the thermal stresses experienced by the components were relatively benign.

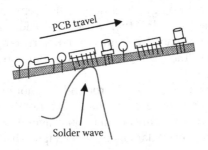

Figure 3.1 Wave soldering through-hole components.

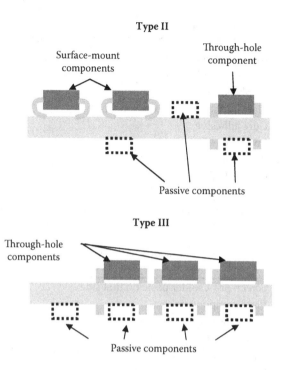

Figure 3.2 A printed circuit board populated with through-hole and surface-mount components (Type II and Type III).

But when PCB technology moved to surface mount, a number of details changed in the solder reflow process. Table 3.2 shows the process steps for mixed assemblies of through-hole and surface-mount packages, also illustrated in Figure 3.2 and generally referred to as a Type II SMT assembly. Boards now had parts mounted on both sides, so it was more difficult to use wave soldering as an attachment process, because there are few, if any, through-holes left. Also, the components would risk being "washed" off by the molten solder, even with the package body attached to the board with heat-resistant adhesive. Instead, solder reflow would be achieved through convection heating, at first through vapor phase ovens and nowadays by infrared (IR) heat. The process flow is shown in Table 3.3. This also meant the packages would experience the same temperature excursions as the leads mounted to the PCB. A PCB with only SMT components is called a Type I SMT assembly and is illustrated in Figure 3.3. In all, the thermal stresses on the components and packages increased considerably with the move to SMT.

Given the increased thermal stresses, the bill of materials (BOM) that had served well through-hole mounting began to fail with the change to

Table 3.2 Typical Attachment Process Flow for Mix of Through-Hole and Surface-Mount Components

1. Insert leaded components
2. Invert printed circuit board
3. Apply adhesive
4. Place surface-mount components
5. Cure adhesive
6. Invert board
7. Wave solder
8. Clean (if necessary)
9. Electrical test

Source: Adapted from Intel Corporation, *Intel® Manufacturing Enabling Guide*, Chapter 1, May 2010, figure 1-2.

Table 3.3 Typical Attachment Process Flow for Surface-Mount Components Only

1. Screen print solder paste on first side
2. Place components
3. Dry solder paste
4. Reflow solder

(If populating a double-sided board, continue to step 5. Otherwise, skip to step 10.)

5. Invert printed circuit board
6. Screen print solder paste on second side
7. Place components
8. Dry solder paste
9. Reflow solder
10. Clean (if necessary)
11. Electrical test

Source: Adapted from Intel Corporation, *Intel® Manufacturing Enabling Guide*, Chapter 1, May 2010, figure 1-3.

Type I

Surface-mount components

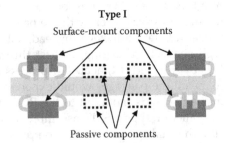

Passive components

Figure 3.3 A printed circuit board populated with surface-mount components only (Type I).

SMT. The main phenomenon associated with the change is called *package cracking* and was nicknamed *popcorning*.

3.4   *Package cracking or "popcorning"*

When semiconductor packages were relatively small, such as small out-line packages, the changeover to SMT did not unduly stress the package integrity. However, when large and thin (3-mm thickness and below) surface-mount packages (such as quad flat packs and plastic leaded chip carriers, and later, plastic ball grid arrays) became prevalent in the late 1980s, the phenomenon nicknamed *popcorning* became well known. Popcorning occurs when a moisture-filled plastic package undergoes the solder reflow process in an in-line oven. The moisture often pools in areas of weakened adhesion, which eventually causes delamination. The package is subjected to rapid heating to temperatures above the boiling point of water. Therefore, the moisture inside the package turns to steam and exerts pressure to escape. The molding compound typically cannot with-stand the force and commonly fails through the backside, resulting in an external crack. The sound of the package failing is said to sound like pop-corn popping, hence the term *popcorning*. The sequence of events is shown in Figure 3.4.

A standard—the IPC-SM-786A—was created by the IPC (the Institute for Interconnecting and Packaging Electronic Circuits) to control mois-ture levels in semiconductor packages prior to solder reflow. The specifi-cation—now superseded by J-STD-020 and J-STD-032—specified storage conditions, the need for dry-packing, or even baking before reflow for very sensitive packages.

Still, there was great interest into the 1990s in discovering chemistries that could make molding compounds that were less moisture sensitive. Research in this area led to the development and use of biphenyl-based molding compounds for large surface-mount packages, which is dis-cussed in further detail in Chapter 5, Section 5.1.

3.5   *Surface-mount packages: peripheral*
leads versus area array

Another difficulty discovered with the transition from through-hole to SMT is that the leads for surface-mount packages are more delicate and fragile than those for through-hole packages. Again, Table 3.1 shows some dimensional comparisons with a plastic quad flat pack—which typically has the highest density, thinnest, and most closely spaced leads of the surface-mount packages—and a plastic dual in-line package.

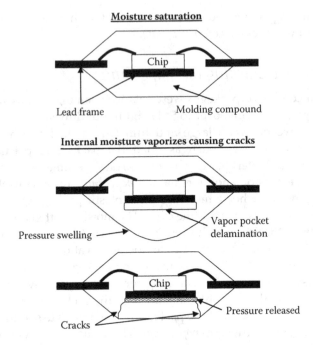

Figure 3.4 Solder reflow package cracking sequence of events.

Unfortunately, the number of I/O (inputs and outputs) on large chips, like microprocessors and field programmable gate arrays (or FPGAs), continues to increase with time and the lead density for packages with peripheral leads soon reached a limit of about 300 leads. Above that number, the leads were simply too delicate to handle in high volume production, not to mention the increased risk of solder bridging and electrical shorts from the closely spaced land pads.

Enter area array packages to fulfill this need, particularly the ball grid array (BGA). As already discussed in Chapter 2, BGAs were developed in the early 1990s and are descended from the through-hole pin grid arrays (PGAs) package family. BGAs appear to be almost miniature PCBs, with an array of solder bumps spaced by a given fixed pitch—hence, the word *grid* in the package's name—on the bottom side, as illustrated in Figure 3.5.

BGA packages, and area array packages in general, have several advantages over surface-mount packages with peripheral leads, beyond that of I/O density. Other advantages include shorter electrical connection distances between chip and main PCB, reducing the likelihood of undesirable electrical characteristics.

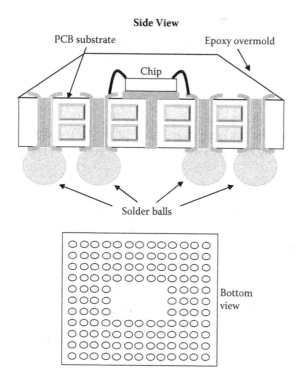

Figure 3.5 Example of a ball grid array package (not to scale).

3.6   Issues with advanced packaging

The development of new and innovative package form factors is often accompanied by changes and difficulties beyond that of package development. As most, if not all, of the new package types coming into production use surface-mount technology, that is one of the hurdles on the way to market acceptance and volume production.

One example of a recent package form factor with particular issues when it comes to surface mounting is package-on-package, or PoP. As shown again in Figure 3.6, a PoP structure is very unbalanced and top

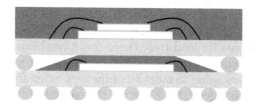

Figure 3.6 Example of a package-on-package (not to scale).

Figure 3.7 Warped package-on-package (not to scale).

heavy, making the package even more prone to warpage during initial package assembly and after solder reflow, as shown in Figure 3.7.

3.7   Current and future trends

Though technology in general tends to always be rapidly changing in the modern age, semiconductor packaging tends to move a bit more slowly, at a more incremental pace. That is not to say changes will not occur, but it does take into account a primary need for reliability out in the field. The longer a semiconductor and its package will be required to be in service or how critical the application is, the more conservative the technology. Good examples are the types of semiconductor packages used in airplanes or automobiles compared to those in a mobile handset or television.

Nonetheless, the industry is subject to external demands and regulations that are pushing for more rapid changes in some cases.

3.7.1   Lead-free and halogen-free packaging

The move to lead (Pb)-free and halogen-free, or *green*, packaging due to environmental, health, and safety regulations—namely the European Union's Restriction of Hazardous Substances or RoHS—has greatly affected the whole surface-mount process. Acceptable solder alloys not containing lead typically require peak reflow temperatures in the range of 30°C to 40°C higher than was needed with eutectic lead-tin solder. This elevated temperature exposure often means the old bill of materials, which worked well with eutectic tin-lead solder, will wither under thermal stresses and will be unable perform reliably, either resulting in infant mortality or field failures.

Slightly different issues are present when eliminating halogens from the bill of materials, but nonetheless, it does means the material set will need to be changed, and it will need to be compatible with the aforementioned requirements for lead-free. Therefore, it has been an ongoing process in the last decade or more to update the bill of materials creating semiconductor packaging to meet all of the new rules and regulations.

In addition, changes extend beyond the package, to the circuit board. As printed circuit boards must also meet the new criteria, they have undergone changes of their own, from new, nonoxidizing finishes on the land pads instead of eutectic lead-tin solder plating to changes in the composition and appearance of the solder mask to the type and composition of the laminate material. As one example, green has been a popular color for solder masks, but it turns out that to achieve that color, the use of materials with unacceptable levels of halogens, especially chlorine, is often required. Another is the various options of finishes and coatings to prevent oxidation of lands, whether an organic solderability preservative (OSP) or using the combination of electroless nickel plating followed by immersion gold coating, abbreviated as ENIG.

So, compatibilities must be found throughout the different levels of electronics assembly, from the semiconductor package to the boards to the enclosure to meet these recent environmental regulations.

Bibliography

A.S. Chen, W.J. Schaefer, R.H.Y. Lo, and P. Weiler, "A Study of the Interactions of Molding Compound and Die Attach Adhesive, with Regards to Package Cracking," presented at 44th ECTC, Washington, DC, 115–120, May 1–4, 1994.

C.A. Harper, *Electronic Packaging and Interconnection Handbook*, McGraw-Hill Professional, New York, Chapter 6, 1991.

Intel Corporation, *Intel® Manufacturing Enabling Guide*, Chapter 1: Component Surface Mount Technology (SMT), May 2010.

Intel Corporation, *2000 Packaging Databook*, Chapter 7: Leaded Surface Mount Technology (SMT), 2000.

L.T. Nguyen, R.H.Y. Lo, A.S. Chen, and J.G. Belani, "Molding Compound Trends in a Denser Packaging World II: Qualification Tests and Reliability Concerns," *IEEE Transactions on Reliability*, vol. 42, no. 4, 518–535, December 1993.

R.D. Skinner, Ed., *Basic Integrated Circuit Technology Reference Manual*, Integrated Circuit Engineering Corporation, Section 3: Packaging, 1993.

V. Solberg, "Designers Guide to Lead-Free SMT," *IPC APEX EXPO*, Las Vegas, NV, March 29–April 2, 2009.

B. Wettermann, "PoP Rework: Process Control and Using the Right Materials Increases Yield," *Advanced Packaging*, September 13, 2010.

What Is SMT Surface Mount Technology, www.radio-electronics.com/info/data/smt/what-is-surface-mount-technology-tutorial.php

chapter four

Other packaging needs

4.1   Objectives

- List and categorize other devices and technologies that require some form of electronic packaging.
- Discuss current status and future developments for these other devices and technologies.

4.2   Introduction

This chapter looks at electronic packaging needs for specialized functions and devices, as well as specialized semiconductor technologies.

4.3   Tape automated bonding

Tape automated bonding (TAB) is an alternate method to wire bonding for creating interconnections from chip to lead frame or substrate.

Introduced into production during the mid-1970s, TAB saw its peak usage in the 1980s to early 1990s, when the fine pitch capability in gold thermosonic wire bonding was still lacking, and overall throughput speeds for wire bonding were still relatively slow. In TAB, the chips and their respective bond pads are attached simultaneously—which is dubbed *gang bonding*—to a continuous tape carrier, which becomes a temporary support for all subsequent manufacturing steps. Prior to gang bonding, either the bond pads or the lead tips receive metallic bumps, typically plated gold.

Table 4.1 describes the TAB process steps. After the wafer is sawed up into individual chips, each chip is bonded to a reel of flexible, polyimide-based tape, with the metal leads cantilevered over the central hole in the tape. The leads are all bonded together into the inner leads. Figure 4.1 shows how the inner TAB leads are now connected to the chip. Next, the chip is connected to a lead frame or substrate via gang bonding the other end, or outer lead bonding.

Table 4.1 Tape Automated Bonding (TAB) Processing
Steps

Key Steps	TAB Process
A	Singulate wafer into individual chips
B	Reel of flexible tape
C	Chip mounted on tape side—inner lead bonding
D	Electrical test on tape
E	Chip bonding to lead frame—outer lead bonding
F	Chip bonding to substrate—outer lead bonding

Source: Adapted from Karel Kurzweil, *Electrocomponent Science and Technology*, 6, 159–163, 1980.

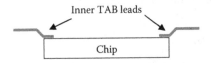

Inner TAB leads

Chip

Figure 4.1 Inner lead tape automated bonding (TAB) (not to scale).

4.4   *Micro electro-mechanical systems (MEMS)*

MEMS (Micro Electro-Mechanical Systems) devices are packaged in a large variety of ways due to the great variation in requirements. These requirements, and the resulting package solutions, go beyond those imposed on semiconductor packaging and result in an unusual variety of packages. For example, MEMS are used for vacuum and atmospheric control, as optical devices, as electrostatic discharge (ESD) devices, for fluid analyzers, and the list goes on. Note that many devices considered as MEMS actually have no moving mechanical parts.

The first choice to package a MEMS device is a standard, off-the-shelf semiconductor package. For example, from STMicroelectronics, a standard 16-lead quad flat no-lead package is used for a three-axis digital output MEMS accelerometer and motion sensor that is known to have been used in the iPhone 3G iteration. Indeed, all of STMicroelectronics' accelerometers and gyroscopes are packaged in quad flat no-lead (QFN) types, ranging in sizes as small as 3.0 by 3.0 by 0.9 mm to as large as 4.4 by 7.5 by 1.0 mm.

The motion sensors generally avoid issues with overmolding "gumming up" the mechanical parts by capping that area to make a protective cavity. With a cap, a MEMS chip simply looks like a stacked-die configuration. Other MEMS device types may require more adaptations or deviations from standard semiconductor (plastic) packaging processes. MEMS for optical applications or with mirror arrays (digital light processors

[DLPs]) often require a cavity package and an opening on top to expose the MEMS section to the visible world. Something similar is likely needed for MEMS microphones, which must ensure that sounds reach the MEMS device clearly and unimpeded. A key emerging technology used to reduce cost and improve performance of MEMS devices is the integration of MEMS devices with standard semiconductor chips, which provide drive, control, and signal processing functions. This approach enables increased integration and reduction in cost.

MEMS devices are sometimes designed to use wafer-level packaging (WLP). Again as an example, an all-digital MEMS microphone on a single complementary metal oxide semiconductor (CMOS) chip by Akustica (acquired by Bosch in the summer of 2009) is in a four-lead WLP. Some wafer-level methods use overmolding; some use wafer-to-wafer bonding; and some build a cavity within the MEMS structure and seal it at the device fabrication level, often with another piece of silicon.

However, in many cases, standard package offerings often end up being inadequate for specific MEMS applications, so engineers must modify the design or design a unique package for manufacturing. The selection and design of a MEMS package can be a major portion of the effort needed to bring a MEMS product to market.

Table 4.2 shows some of the packaging methods used for MEMS, and Table 4.3 presents MEMS packaging examples.

4.5   Image sensor modules

An image sensor package is an optical element, whether based on a complementary metal oxide semiconductor (CMOS) or charge coupled device (CCD) packaged with an optical lens that enables the imaging functions.

Packaging needs for digital image sensors have their own special requirements. For one thing, the sensors need to "see," so nothing opaque can cover the chip surface. Also, the sensor surface and the clear cover must be kept extremely clean and handled carefully, so that dust and scratches do not degrade the image capture quality. Image sensor packages may come in many different formats, all derived from established package types. One of the more commonly used ones is the LCC (leadless chip carrier); an example using a plastic molded body with an opening on the top surface for the image sensor is shown in Figure 4.2. Images sensors packaged inside a LCC are cavity-up, wire bonded, and given a glass lid. They offer a small form factor and excellent thermal performance.

4.6   Memory cards

The advent of the digital still camera brought about the initial growth and demand for compact, removable memory cards. Unfortunately, at

Table 4.2 Package Types Used for Micro Electro-Mechanical System (MEMS) Devices

Type of Package	Applications	Characteristics
Plastic overmolded, leaded and leadless	Resonators, accelerators, inertial sensors	Low cost
Premolded plastic air cavity, leaded and leadless	Pressure sensors, accelerators, microphones	Low-cost cavity package
Ceramic with metal lid or metal cap	Lab-on-a-chip, optical devices, radio-frequency (RF) switches	Highly stable, costly, complex to engineer and fabricate; control of cavity environment; dry, vacuum, inert gas, and so forth
Ceramic with glass cover	Optical applications, charge-coupled device (CCD) packages, digital light processor (DLP)	Stable, moderate cost, optical window
TO-5 with hole or window	Pressure sensors, some optical devices	Low cost, widely available
Glass-on-glass	Optical applications, displays	Large cavity packages, sometimes with stand-offs
MEMS on organic substrates with glass cover	Optical switches, displays	Quick to market, low cost
MEMS on substrate— organic, ceramic, and so forth—wire bond, partial, or total encapsulation	Ink-jet print heads, fingerprint readers	
Wafer level (structure built, then singulate)	Camera modules	Lowest cost in volume production

Source: Adapted from *International Technology Roadmap for Semiconductors, 2007 Edition,* Assembly and Packaging chapter.

least for the consumer, product development left to free market forces usually means a proliferation of many incompatible formats. And so was the case with memory cards—at recent count, there are approximately 14 types, including variations on a given basic form factor. Figure 4.3 shows a secure digital (SD) card, which is only one of the many different kinds of memory card formats currently available on the market. Table 4.4 lists most of the available formats and their specifications.

Table 4.3 Micro Electro-Mechanical Systems (MEMS) Packaging Examples

Market	Automotive	Consumer	Two-Dimensional (2D) Optical Switch	Three-Dimensional (3D) Optical Switch	Network Switch	Wireless
Application	Acceleration, airbag sensor	Video games, appliances	Optical add-drop multiplexer (OADM)	Wide-area network (WAN) and local area network (LAN)	Electronic switches	Surface acoustic wave (SAW) filters
MEMS type	Two-axis accelerometer	Three-axis accelerometer	64 mirrors, 90° motion	180° full motion mirrors	Contact switch	Planar filter
Package size	TO-8, 14L CerDIP	Surface-mount ceramic leaded chip carrier (CLCC)	Custom metal, 82 mm²	Custom ceramic, 184 mm²	Low-temperature co-fired ceramic (LTCC), 27 mm²	Printed wiring board (PWB), 40 mm²
Clean room	100; 10,000	100; 10,000	100; 10,000	100; 10,000	100	10,000
Die bond	Epoxy	Epoxy	Au-Sn Eutectic	Epoxy	Au-Sn Eutectic	Epoxy
Wire bond	0.7–1.0 Au ball	0.7–1.0 Au ball	1.25 Al, 1.25 Au	1.25 Au wedge/ball	1.0–1.25 Au ball	1.0 Au ball
Seal	Seam seal	Molded	Seam seal	Seam seal	Epoxy	Epoxy lid seal
Leak test	Gross/fine	None	Gross/fine	Gross/fine	Gross/fine	None
Additional	N/A	N/A	Fiber optics, connectors	Flex circuit, connectors	PWB connectors	SMT connectors
Manufacturing level	Production	Production	Preproduction	Preproduction, research and development (R&D)	Preproduction, prototype	Preproduction, prototype

Source: Adapted from *International Technology Roadmap for Semiconductors, 2007 Edition*, Assembly and Packaging chapter.

Figure 4.2 Image sensor package. (From Wikimedia Commons.)

Figure 4.3 Secure digital (SD) memory card. (From WP Clipart.)

Table 4.4 Memory Card Formats

Card Format	Size (mm)	Total Interface Pin Count	Data Interface Pin Count	Interface Clock Rate (MHz)	Maximum Data Rate at Host Interface (Mbytes/sec)	Voltage (V)	Built-In Memory Controller	Security
PCMCIA	85.6 × 53.8 × 3.3 (Type I); 85.6 × 53.8 × 5 (Type II); 85.6 × 53.8 × 10.4 (Type III)	68	8, 16, or 32	16 (PC Card); 132 (CardBus)		3.3, 5	No	No
Compact Flash	42.8 × 36.4 × 3.3 (Type I); 42.8 × 36.4 × 5 (Type II)	50	8 or 16	N/A	16	3.3, 5	Yes	No
Smart Media	37 × 45 × 0.8	22	8	20	20	1.8, 3.3	Yes	Optional
Multimedia card (MMC)	24 × 32 × 1.4	7	1	20	2.5	1.8, 3.3	Yes	Optional
RS-MMC	24 × 18 × 1.4	7	1	20	2.5	1.8, 3.3	Yes	Optional
MMC plus	24 × 32 × 1.4	13	8	52	52	1.8, 3.3	Yes	Optional
MMC mobile	24 × 18 × 1.4	13	8	52	52	1.8, 3.3	Yes	Optional
Secure Digital (SD)	24 × 32 × 2.1	9	4	20	10	3.3	Yes	Yes
Mini SD	20 × 21.5 × 1.4	9	4	20	10	3.3	Yes	Yes
Micro SD	10 × 15 × 1.1	9	4	20	10	3.3	Yes	Yes
Memory Stick	21 × 50 × 2.8	10	1	24	3	3.3	Yes	No
Memory Stick Duo	20 × 31 × 1.6	10	1	24	3	3.3	Yes	No
Memory Stick Pro	20 × 50 × 2.8	10	4	40	20	1.8, 3.3	Yes	Yes
Memory Stick Pro Duo	20 × 31 × 1.6	10	4	40	20	1.8, 3.3	Yes	Yes
xD-Picture Card	25 × 20 × 1.7	18	8	N/A	20	3.3	Yes	No
USB Flash Drive	Various	4	1 (differential)	12 (full speed); 480 (high speed)	1.5 (full speed); 60 (high speed)	5	Yes	Optional

Source: Adapted from Brian Dipert, *EDN*, 53–61, July 8, 2004.
Note: PCMCIA, Personal Computer Memory Card International Association cards; RS, reduced size.

After digital still cameras, the next market memory card conquered was mobile phones. As handsets morphed into camera-phones and smart-phones, the need for extra memory capacity became readily apparent, and soon mobile phones with slots for removable memory cards were ubiquitous. Most handsets use a smaller version of the SD card format, either mini-SD or the even smaller micro-SD.

Memory Stick is a format developed by Sony, and generally their products are the only ones using that form factor. The xD-Picture card is typically found only in digital cameras.

Even TVs and printers now often sport memory card slots in the front, for directly downloading data and images.

The number of formats seems to be stabilizing now, and many of the formats are falling into disuse or have become obsolete for various reasons (size, memory capacity limitations, etc.). Among those no longer or rarely used include Memory Stick, xD-Picture Card, and Smart Media. Currently, the two card formats commonly found in retail outlets are SD and its various micro and mini variations, and CompactFlash (CF). SD is the more dominant, supported in a wide range of products from digital cameras to laptops to smartphones to televisions. CF is generally only found in high-end SLR (single-lens reflex) digital cameras and other specialty applications. These two card formats are likely to remain in the marketplace for some time, as both have upcoming transfer speed and memory capacity improvements in the works by their respective standard bodies.

4.7   Packaging needs for solar technology

Traditional semiconductor package technologies are not employed in packaging solar cells and solar cell arrays. However, there are unique packaging requirements to package solar cells and solar cell arrays against the elements.

Solar cell modules face temperature extremes and must have a very long life compared to almost any other semiconductor packaging requirement. The current state of the art for the photovoltaic modules used in solar cell arrays is as follows:

- Semiconductor thickness 180 μm
- Soldered with high-throughput tabber-stringer
- Vacuum lamination
- EVA (ethylene-vinyl acetate) as encapsulant
- Guaranteed lifetime of 25 years

The continued expansion of solar power and the changes anticipated in solar cell technology and operational demands will eventually move to more stringent requirements. These new requirements will include the following:

- Low-stress interconnection for very thin solar cells (between 100- and 150-μm thick)
- High-throughput lamination technology
- Lead-free soldering solutions
- 30-years lifetime
- Design for easy recycling at end of life

Semiconductor packaging only comes directly into play with solar collection systems. Semiconductors are required to regulate voltages and current, as well as DC-to-AC power inverters and other power management purposes. These power management chips come in standard package form factors and do not have extraordinary requirements specific to solar power configurations.

Bibliography

Akustica, Inc., www.akustica.com

"CMOS Image Sensor Packaging: Picture the Challenges," *Prismark Partners LLC*, November 1999.

C. Crossman, "Device Reads, Writes in 12 Formats," *The Seattle Times*, March 21, 2005.

B. Dipert, "Pick a Card," *EDN*, 53–61, July 8, 2004.

B. Dipert, "Flash Forward to the Future," *EDN*, November 1, 2004.

Y. Emoto, N. Ohikata, Y. Kawakami, M. Konda, T. Yamamoto, K. Chiba, and N. Kunii, "Development of Molded TAB Package Technology," *Nippon Steel Technical Report*, no. 56, 1–6, January 1993.

B. Howard, "Flash Memory: Pick a Card," *PC Magazine*, September 2, 2003.

International Technology Roadmap for Semiconductors, 2007 Edition, Assembly and Packaging chapter.

R.C. Johnson, "MEMS Mics Moving into Mainstream," *EE Times*, September 22, 2008.

R.C. Johnson, "Bosch Acquires MEMS Microphone Pioneer Akustica," *EE Times*, August 19, 2009.

K. Kurzweil, "An Installed Tape Automated Bonding Unit," *Electrocomponent Science and Technology*, vol. 6, 159–163, 1980.

K. Kurzweil, "Tape Automated Bonding for High Density Packaging," *Electrocomponent Science and Technology*, vol. 8, 15–19, 1981.

R.H.Y. Lo and E. Tjhia, "Backsputtering Etch Studies in Wafer Bumping Process," *Solid State Technology*, pp. 91–94, June 1990.

R. Lo, "Upcoming Trends in IC Package Technology," *SEMICON Taiwan 2001*, September 17–19, 2001.

R. Lo and H.P. Takiar, U.S. Patent No. 5,617,297: Encapsulation filler technology for molding active electronics components such as IC cards or PCMCIA cards, April 1, 1997.

SanDisk presentation, *Lehman Brothers T4 2004—Technology and Telecom Trends for Tomorrow*, December 8, 2004.

S. Shankland, "CompactFlash Allies Rally Against Dominant SD," *CNET News*, December 14, 2010.

Siliconware Precision Industries Ltd., www.spil.com.tw/

STMicroelectronics, Data Sheet, *LIS221DL*, April 2008.

STMicroelectronics, Product Brochure, *MEMS Motion Sensors*, October 2009.

"STMicroelectronics Rolls Out Nano Three-Axis Linear Accelerometers," *EDA Geek*, October 2, 2007.

section two

Package reliability

chapter five

Reliability testing

5.1 Introduction

Once the chip has been packaged and electrically tested, there is the requirement that the package demonstrate field reliability. Because it is impossible to assess that in real time and on all device/package combinations out in the field, accelerated testing is done on a statistically significant sample size to confirm operational reliability.

Qualification or reliability—accelerated—tests are aimed at the following:

- Inducing typical failure modes rapidly
- Decreasing the amount of time required to assess the reliability of a given package and device combination
- Screening out defects early and quickly
- Forecasting what might be the useful lifetime of a given part based on acceleration factors (This useful life period can be called mean time to failure [MTTF] in the case of nonrepairable semiconductor devices in packages.)

This is illustrated in Figure 5.1, which is called the reliability bathtub curve.

Naturally, these tests are only useful if the failure mechanisms induced reflect those encountered under normal operating conditions. Even though semiconductor packages are generally robust and do their job of protecting the chip, or chips, within, they are nevertheless subject to a variety of failure modes.

Each material component of a package exhibits a typical or common failure mode peculiar to that material or purpose, though the final manifestation may be common across several materials. For instance, underfill materials can crack during temperature or power cycling, either during reliability testing or during actual field operation. The thermal expansion mismatch between the chip and underfill causes stress concentrations in certain locations, like at a die corner. As the crack propagates during continued stressing, the crack opens up the chip–underfill interface or another weakened section, causing either interconnect fatigue or an electrical failure in the chip's dielectric.

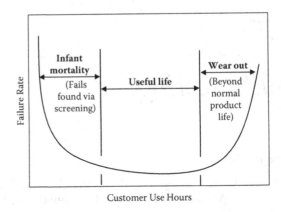

Figure 5.1 Reliability bathtub curve for a device's useful life (not to scale).

5.2   Background

Standards created by industry organizations help determine the number and type of tests, as well as test conditions, that best cover the possible failure mechanisms in a consistent manner across all suppliers in the industry. The original standard developed and used by the semiconductor industry was Mil-Std-883, which was used to qualify semiconductor devices and packages for military and aerospace applications, but in those early days, most, if not all, of the packaging done was hermetic, with chips sealed inside metal cans or sandwiched between ceramic substrates and caps.

Later, the industry association JEDEC (Joint Electron Devices Engineering Council, now officially known as *JEDEC Solid State Technology Association*)—mentioned in Chapter 2 as the standards body for the semiconductor industry—created testing standards applicable for plastic semiconductor packages. The most current set of reliability and qualification criteria fall under the JESD22 family of test methods.

5.3   Examples of reliability tests

There are many different reliability tests, each aimed at eliciting a certain failure mode. There is also some overlap between various tests, as some are considered too harsh to mimic more benign operating conditions (i.e., a desktop PC in an office environment versus a ruggedized laptop for military operations or an embedded computer module under the hood of an automobile). The same goes for the test conditions used. Also, the semiconductor suppliers can pick and choose the types and specifications of the tests and vary them away from industry standards, as per the needs and requirements of their end customers. Some of the important tests are listed in Table 5.1.

Table 5.1 Partial List of Reliability Tests

Test	Abbreviation	Goal	Test Conditions
Preconditioning	Precon	Mimics solder reflow board attachment for surface mount plastic packages; done prior to other reliability tests	See Table 5.2
High-temperature storage	HTS	Induce bond pad metallization failures through heat and halides	1000 hours at 150°C or 175°C
Temperature cycling	TC	Thermal stressing of the physical construction of the chip and package	−65°C to 150°C for 1000 cycles in air
Thermal shock	TS	Thermal stressing of the physical construction of the chip and package	−65°C to 150°C for 100 to 1000 cycles in liquid
Temperature-humidity-bias	THB	Passivation integrity against ionics under moisture and electrical bias conditions	1000 hours at 85°C/85% RH (relative humidity)
Autoclave	ACLV	Passivation integrity against ionics under moisture conditions	168 hours or more at 121°C and between 15 and 30 psig
Highly accelerated temperature and humidity stress testing	HAST	An accelerated version of the THB test	110°C or 130°C under 85% RH and under vapor pressure for at least 96 or more hours

Source: Adapted from JEDEC, *JEP150, Stress-Test-Driven Qualification of and Failure Mechanisms Associated with Assembled Solid State Surface-Mount Components*, May 2005, table 1.

There are JEDEC standards (for instance, JESD74) to help determine a statistically significant sample size, which should be drawn from, and divided into, a minimum of three nonconsecutive production lots, to help determine early life failure rates (ELFRs).

5.3.1   Preconditioning conditions

The requirements can vary considerably, depending on a given package form factor's sensitivity to moisture or moisture sensitivity level (MSL) level, which is detailed in the IPC/JEDEC J-STD-020 document. The table detailing MSL levels in shown in Table 5.2. The current industry criteria are laid out in Table 5.3, which shows the typical sequence of events. The solder reflow profiles specified for a given package form factor are given in Table 5.4 and illustrated in Figure 5.2.

As stated in the title of JESD22-A113F, the preconditioning test is done prior and in addition to other reliability tests. In other words, parts set aside to under temperature cycling would undergo a preconditioning sequence before starting the temperature cycling test sequence. Preconditioning is not intended as a stand-alone reliability or qualification test.

Preconditioning is intended to simulate the viability of the packages during board assembly and mimic the environment of the production floor, where packages might sit out in the open, away from their dry-pack bags or from a dry box, and be saturated with moisture.

As already discussed in the chapters for molding compounds and for interfacial interaction, the goal of preconditioning is to see whether the package-die system can resist the phenomenon of solder reflow package cracking, otherwise named *popcorning*.

5.3.1.1   Package failure mode: package crack or popcorning

Indirect failures from moisture include package popcorn cracking of large surface-mount packages during solder reflow. As an aside, the term *popcorn cracking* refers to the often-audible fracture of the interface between molding compound and chip or its lead frame or substrate, commonly due to insufficient interfacial adhesion, especially if sufficient moisture is present. Figure 5.3 shows the various interfaces where delamination can occur and stress concentration areas develop for cracks to propagate. Figure 5.4 illustrates the sequence of events leading to package failure. Even without the appearance of an external crack, the package's integrity is almost certainly degraded from delamination between internal interfaces.

5.3.2   Temperature cycling and thermal shock

The purpose of temperature cycling is to check on package integrity in spite of the coefficient of thermal expansion mismatches between the

Table 5.2 Preconditioning Test Conditions

Level	Floor Life Time	Soak Requirements Condition	Standard Time (hours)	Condition	Accelerated Equivalent[a]		
					eV 0.40–0.48 Time (hours)	eV 0.30–0.39 Time (hours)	Condition
1	Unlimited	≤30°C/85% RH	168 +5/–0	85°C/85% RH	N/A	N/A	N/A
2	1 year	≤30°C/60% RH	168 +5/0	85°C/60% RH	N/A	N/A	N/A
2a	4 weeks	≤30°C/60% RH	696[b] +5/0	30°C/60% RH	120 +1/0	168 +1/–1	60°C/60% RH
3	168 hours	≤30°C/60% RH	192[b] +5/0	30°C/60% RH	40 +1/0	52 +1/0	60°C/60% RH
4	72 hours	≤30°C/60% RH	96[b] +5/0	30°C/60% RH	20 +1/0	24 +1/0	60°C/60% RH
5	48 hours	≤30°C/60% RH	72[b] +5/0	30°C/60% RH	15 +1/0	20 +1/0	60°C/60% RH
5a	24 hours	≤30°C/60% RH	48[b] +5/0	30°C/60% RH	10 +1/0	13 +1/0	60°C/60% RH
6	Time on label (TOL)	≤30°C/60% RH	TOL	30°C/60% RH	N/A	N/A	N/A

Note: Suppliers may extend the soak times at their own risk.

Source: Adapted from JEDEC, IPC/JEDEC J-STD-020D.1, *Moisture/Reflow Sensitivity Classification for Nonhermetic Solid State Surface Mount Devices*, March 2008, table 5-1.

[a] Caution: To use the "accelerated equivalent" soak conditions, correlation of damage response (including electrical, aftersoak, and reflow) should be established with the "standard" soak conditions. Alternately, if the known activation energy for moisture diffusion of the package materials is in the range of 0.40 to 0.48 eV or 0.30 to 0.39 eV, the "accelerated equivalent" may be used. Accelerated soak times may vary due to material properties (e.g., mold compound, encapsulant, etc.). JEDEC document JESD22-A120 provides a method for determining the diffusion coefficient.

[b] The standard soak time includes a default value of 24 hours for semiconductor manufacturer's exposure time (MET) between bake and bag and includes the maximum time allowed out of the bag at the distributor's facility. If the actual MET is less than 24 hours, the soak time may be reduced. For soak conditions of 30°C/60% RH, the soak time is reduced by 1 hour for each hour the MET is less than 24 hours. For soak conditions of 60°C/60% RH, the soak time is reduced by 1 hour for each 5 hours the MET is less than 24 hours. If the actual MET is greater than 24 hours, the soak time must be increased. For soak conditions of 30°C/60% RH, the soak time is increased 1 hour for each hour the actual MET exceeds 24 hours. For soak conditions of 60°C/60% RH, the soak time is increased 1 hour for each 5 hours the actual MET exceeds 24 hours.

Table 5.3 Steps in Conducting Preconditioning Testing

Step	Item	Details
1	Initial electrical test	Replace any failing devices; optional for testing by supplier
2	Visual inspection	Replace any failing devices; optional for testing by supplier
3	Temperature cycling	Five cycles at –40°C to 60°C; optional shipping simulation based on product requirements
4	Bake	24 hours at 125°C; optional for testing by supplier
5	Moisture soak	Soak time and conditions per IPC/JEDEC J-STD-020 based on device MSL level
6	Reflow	Three reflow cycles using profiles per IPC/JEDEC J-STD-020, document rev of J-STD-020 used; SnPb or Pb-free profile based on device and use process
7	Flux application	10 seconds of full immersion in activated water-soluble flux; optional for testing by user or second-level configuration; not required for ball grid array (BGA), column grid array (CGA), and land grid array (LGA) packages
8	Cleaning	Deionized (DI) water rinse; remove all flux residual; optional for testing by user or second-level configuration; not required for BGA, CGA, and LGA packages
9	Drying	Room ambient drying; optional for testing by user or second-level configuration; not required for BGA, CGA, and LGA packages
10	Final electrical test	If all devices pass, then ready for reliability testing; if valid failures are found, then devices may have been tested to the wrong MSL level or something is substandard with the devices; optional for testing by supplier

Source: Adapted from JEDEC, *JESD22-A113F, Preconditioning of Plastic Surface Mount Devices Prior to Reliability Testing*, October 2008, Annex A.

various materials and their respective interfaces within the plastic package. Unlike hermetic packages, the die surface is not isolated from the other material surfaces within the package. Good contact and adhesion are key to a long service life. All the components—the semiconductor chip, the metal lead frame or organic substrate, the polymer-based molding compound or die attach adhesive—within a package will likely have very different expansion and contraction rates while being heated or otherwise

Table 5.4 Solder Reflow Profiles

Condition	Sn-Pb Eutectic Assembly	Pb-Free Assembly
Preheat/soak		
Minimum temperature	100°C	150°C
Maximum temperature	150°C	200°C
Time from minimum to maximum	60 to 120 seconds	60 to 120 seconds
Ramp-up rate	3°C per second maximum	3°C per second maximum
Liquidous temperature (T_L)	183°C	217°C
Time held above T_L	60 to 150 seconds	60 to 150 seconds
Peak package body temperature	Between 220°C and 235°C, depending on size and volume	Between 245°C and 260°C, depending on size and volume
Hold time within 5°C of peak temperature	20 seconds	30 seconds
Ramp-down rate	6°C per second maximum	6°C per second maximum
Time from 25°C to peak temperature	6 minutes maximum	8 minutes maximum

Source: Adapted from JEDEC, *IPC/JEDEC J-STD-020D.1, Moisture/Reflow Sensitivity Classification for Nonhermetic Solid State Surface Mount Devices,* March 2008, table 5-2.

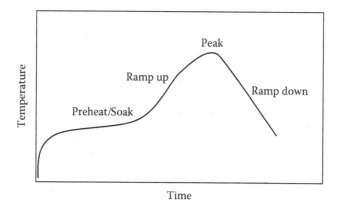

Figure 5.2 Solder reflow profile over time and temperature (not to scale).

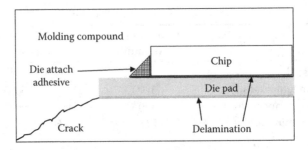

Figure 5.3 Package material interfaces prone to delamination and subsequent package cracking (not to scale).

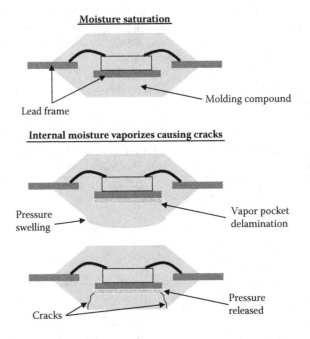

Figure 5.4 Sequence of events during package cracking, or *popcorning*, event.

thermally stressed, as already illustrated in Table 1.3 and shown again in Table 5.5. Cycling between expansion and contraction causes movement between the plastic, epoxies, metal lead frame, and chip. Because the chip expands much less than either plastic or metal, its surface—along with all of the delicate active circuitry and dielectric layers—is prone to damage. The typical failure modes seen with temperature cycling are delamination between interfaces or cracking somewhere in the die, whether it be the passivation, the interlayer dielectric, or through the bulk silicon. Therefore, the chip metallization may see movement out of position.

Table 5.5 Key Properties of Semiconductor Packaging Materials

Material	Coefficient of Thermal Expansion (CTE) (ppm/°C)	Density (g/cm³)	Thermal Conductivity (W/m*K)	Electrical Resistivity (μΩ-cm)	Tensile Strength (GPa)	Melting Point (°C)
Silicon	2.8	2.4	150	N/A	N/A	1430
Molding compound	18–65	1.9	0.67	N/A	N/A	165 (T$_g$)
Copper	16.5	8.96	395	1.67	0.25–0.45	1083
Alloy-42	4.3	N/A	15.9	N/A	0.64	1425
Gold	N/A	19.3	293	2.2	N/A	1064
Aluminum	23.8	2.80	235	2.7	N/A	660
Eutectic tin-lead solder	23.0	8.4	50	N/A	83	183
Alumina	6.9	3.6	22	N/A	N/A	2050
Aluminum nitride	4.6	3.3	170	N/A	N/A	2000

Source: Adapted from National Semiconductor Corporation, *Data Sheet: Semiconductor Packaging Assembly Technology,* August 1999.

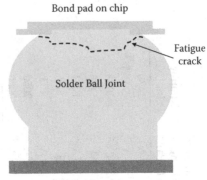

Bond pad on chip

Fatigue
crack

Solder Ball Joint

Bonding land on substrate

Figure 5.5 A ductile solder fatigue-induced crack due to temperature cycling (not to scale).

5.3.2.1   *Package failure modes from temperature cycling and thermal shock*

Temperature cycling and thermal shock are the cause of numerous failure modes in semiconductor packages. Some of these modes include broken bond wire or lifted bonds from pads, solder joint/bump/ball fatigue (an example of which is shown in Figure 5.5), cracked molding compound, and the aforementioned interface delamination. Related to these issues is the thermal expansion coefficient and Young's modulus mismatches.

5.3.2.2   *Package failure mode: delamination*

Of the failure modes mentioned, delamination is a key concern, as it can aid corrosion by creating one or more pathways for moisture ingress. It is further discussed in Section 5.3.4.

Delamination—or the loss of adhesion between interfaces—is often a first step toward a reliability failure, whether it be popcorning or bond wire lift or cracked silicon. The causes are varied—molding compound or underfill shrinkage, surface contamination, and thermal stresses, to name a few.

With die attach adhesive, several failure modes are the result of voids or delamination in the interface with the chip or with the lead frame or substrate pad. Voids can result in hot spots that could cause overheating or thermal breakdown of the die. Delaminated areas may allow for stress concentrations, which could result in cracked chips or cracks to grow into the molding compound. Separation is often caused by thermal expansion mismatch stresses along the dissimilar material interfaces.

Overall, the remedies are not so simple. Enhancing adhesion is not always straightforward and must be balanced against other desired properties, similar to the issue of corrosion and its possible solutions.

5.3.3   High-temperature storage life

The intent of the high-temperature storage test is to elicit premature bond failure, whether in wire bonds or in flip-chip bumps. Elevated temperatures are known to accelerate the typical failure mechanisms, which usually involve growth of brittle intermetallic phases.

5.3.3.1   Package failure mode: intermetallics

In the case of gold wire bonding, it is usually excessive intermetallic growth between the gold ball bond and aluminum bond pad, which is discussed further in the chapter covering bonding wires (see Chapter 7, Section 7.2). The intermetallic layer can consume the thin layer of aluminum, due to the Kirkendall effect, and is relatively porous, so bond resistance increases with time and temperature until an electrical opening occurs. Sometimes this defect is an intermittent one, as repeated biasing of the bad bond will "re-weld" the bond to the pad, at least temporarily.

Gold-aluminum intermetallic growth becomes even more accelerated in the presence of halides. And before the advent of "green" or halogen-free materials, molding compounds would be the most likely source, because both bromine and antimony oxides were used as flame retardants. Now that halogen-free molding compounds will become prevalent, this particular type of failure mechanism will likely disappear.

5.3.4   Temperature-humidity-bias tests

The purpose of subjecting parts to storage at 85°C/85% relative humidity and under electrical bias is to ensure that the devices and packages can withstand an uncontrolled, moisture-laden environment for an extended period of time, perhaps as long as their operating lives. More stressful versions of this test are autoclave and highly accelerated stress test (HAST).

5.3.4.1   Package failure mode: corrosion

Corrosion is the primary mechanism for failure with these types of tests. The phenomenon of corrosion is more easily seen in a humid environment than in a dry one. Thus, it is the common failure mechanism seen in autoclave, temperature-humidity (and perhaps with bias) and HAST test methods. What is typically observed is moisture-induced corrosion of aluminum bond pads or interconnects due to the presence of ionic impurities like chloride, which is shown in Figure 5.6. The ionic impurities travel to the semiconductor surface through diffusion and moisture penetration under these harsh test conditions.

Given that plastic packages are nonhermetic, moisture absorption will always be unavoidable and corrosion will always be a risk. Preventing

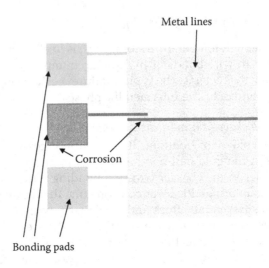

Figure 5.6 Aluminum bond pad corrosion (not to scale).

corrosion in plastic packages needs a multipronged approach. Naturally, reducing the amount of ionic impurities will help, but it is impossible to remove all impurities from packaging materials, and that does not take into account any instance of accidental contamination through human or other errors. Incorporating ion-getters into the materials set may help trap the ions, but care must be taken to ensure the getters do not cause new problems. Altering packaging materials to lower their moisture absorption levels will help, but again care must be taken that other material properties are not altered detrimentally.

5.4   Limitations of reliability testing

The intent of reliability tests is to elicit known failure mechanisms through a set of accelerated tests and provide an accurate level of field life prediction. However, not all failure modes can be accounted for, due to the unpredictable nature of things.

Two kinds of failures are very hard to predict. One is one-off defects, which can come from unknown flaws in the chip or some minor mishap during manufacturing. Examples would be electrostatic discharge and particulate contamination.

Another type of defect that would be hard to detect by reliability testing is an intermittent defect. Marginal wire bond adhesion or flip-chip connections are typical intermittent failures, where the failure may "recover" between testing points. In situ testing is best at catching these kinds of failures.

Bibliography

B. Dipert, "Silicon Contends with Stuffed and Shrinking Packages," *EDN*, pp. 49–58, June 13, 2002.

A. Jalar, M.F. Rosle, and M.A.A. Hamid, "Effects of Thermal Aging on Intermetallic Compounds and Voids Formation in AuAl Wire Bonding," *Solid State and Technology*, vol. 16, no. 2, 240–246, 2008.

JEDEC, *IPC/JEDEC J-STD-020D.1, Moisture/Reflow Sensitivity Classification for Nonhermetic Solid State Surface Mount Devices*, March 2008.

JEDEC, *JEP122E: Failure Mechanisms and Models for Semiconductor Devices*, March 2009.

JEDEC, *JEP148A, Reliability Qualification of Semiconductor Devices Based on Physics of Failure Risk and Opportunity Assessment*, December 2008.

JEDEC, *JEP150, Stress-Test-Driven Qualification of and Failure Mechanisms Associated with Assembled Solid State Surface-Mount Components*, May 2005.

JEDEC, *JEP156: Chip-Package Interaction—Understanding, Identification and Evaluation*, March 2009.

JEDEC, *JESD22-A101C, Steady-State Temperature Humidity Bias Life Test*, March 2009.

JEDEC, *JESD22-A104D, Temperature Cycling*, March 2009.

JEDEC, *JESD22-A110C, Highly Accelerated Temperature and Humidity Stress Test (HAST)*, January 2009.

JEDEC, *JESD22-A113F, Preconditioning of Plastic Surface Mount Devices Prior to Reliability Testing*, October 2008.

JEDEC, *JESD74, Early Life Failure Rate Calculation Procedure for Electronic Components*, October 2000.

JEDEC, www.jedec.org

F.L.A. Latip, A. Hassan, and R. Yahya, "Delamination and Void Analysis on Die Attach Epoxy of a QFN Package," *Solid State Science and Technology*, vol. 16, no. 2, 207–213, 2008.

R.H.Y. Lo and A.S. Chen, "Unconventional Molding Compounds for Conventional Packages," in *Proceedings FOCUS '94 Expo and Conference*, San Jose, CA, August 30–September 1, 1994.

A.F. Moor, A. Casanovas, and S.R. Purwin, "The Case for Plastic-Encapsulated Microcircuits in Spaceflight Applications," *Johns Hopkins APL Technical Digest*, vol. 20, no. 1, 91–100, 1999.

National Semiconductor Corporation, *Data Sheet: Semiconductor Packaging Assembly Techonology*, August 1999.

L.T. Nguyen, R.H.Y. Lo, A.S. Chen, H. Takiar, and J.G. Belani, "Molding Compounds Trends in a Denser Packaging World: Qualification Tests and Reliability Concerns," *IEEE Transactions on Reliability*, vol. 42, no. 4, 518–535, December 1993.

L.T. Nguyen, A.S. Chen, and R.H.Y. Lo, "Interfacial Integrity in Electronic Packaging," *ASME 1995—Application of Fracture Mechanics in Electronic Packaging and Materials*, EEP-vol. 11/MD-vol. 64, 35–44, 1995.

Sony Semiconductor, *Quality and Reliability Handbook, Chapter 2: Failure Mechanisms*, revised May 2001.

Materials used in semiconductor packaging

chapter six

Polymers

6.1 Molding compounds

6.1.1 Objectives

- Describe what a molding compound is.
- Convey its purpose and importance in semiconductor packaging, in regard to physical performance and contribution to reliability assessment.
- Illustrate continuous development and improvement of this material.

6.1.2 Introduction

This chapter briefly reviews the history and use of molding compounds in plastic semiconductor packages, addresses some of the issues and failure modes associated with molding compounds, and touches on future trends.

6.1.3 Background

Molding compounds have been around a very long time, since the advent of the through-hole plastic dual-inline package (DIP) family in the early 1970s. Essentially, molding compounds are epoxy resins filled with some sort of silica filler to reduce the coefficient of thermal expansion to better match that of the lead frame, along with small amounts of other additives, such as carbon black for color and bromine to act as a flame retardant. A comparison of the physical properties for the various components that constitute a plastic package was previously given in Table 1.3 and is repeated as a refresher in Table 6.1.

Initially, the predominant epoxy compound used was bisphenol-A. Epoxy cresol novolac replaced bisphenol-A as the preferred epoxy resin due to its better heat resistance. In general, epoxy resins became the preferred backbone for molding compounds due to their inherent low viscosity, fast cure properties, low shrinkage during cure, good adhesion to the other components in a chip package, and good overall mechanical stability. Both epoxy cresol novolac and bisphenol-A produce sodium chloride as by-products during synthesis. Because both elements are detrimental

Table 6.1 Key Properties of Semiconductor Packaging Materials

Material	Coefficient of Thermal Expansion (CTE) (ppm/°C)	Density (g/cm³)	Thermal Conductivity (W/m*K)	Electrical Resistivity (µΩ-cm)	Tensile Strength (GPa)	Melting Point (°C)
Silicon	2.8	2.4	150	N/A	N/A	1430
Molding compound	18–65	1.9	0.67	N/A	N/A	165 (T_g)
Copper	16.5	8.96	395	1.67	0.25–0.45	1083
Alloy-42	4.3	N/A	15.9	N/A	0.64	1425
Gold	N/A	19.3	293	2.2	N/A	1064
Aluminum	23.8	2.80	235	2.7	83	660
Eutectic tin-lead solder	23.0	8.4	50	N/A	N/A	183
Alumina	6.9	3.6	22	N/A	N/A	2050
Aluminum nitride	4.6	3.3	170	N/A	N/A	2000

Source: Adapted from National Semiconductor Corporation, *Data Sheet: Semiconductor Packaging Assembly Technology,* August 1999.

to integrated circuit (IC) reliability, care must be taken to remove them from the final resin product before molding compound formulation.

The filler comes in the form of amorphous or crystalline silica. Sometimes alumina is used as the filler for increased thermal conductivity and high heat dissipation properties, but it is very abrasive compared to silica. Amorphous silica is preferred when a low thermal expansion coefficient is needed, and crystalline silica provides some thermal conductivity at the expense of a higher coefficient of thermal expansion. In an epoxy cresol novolac–based compound, the filler makes up 65% to 75% by weight, with the resin constituting the majority of the balance. Fillers provide mechanical strength to the compound and reduce the thermal expansion coefficient, which, in turn, reduces shrinkage after molding. Fillers do have one major risk, in that silica may contain minute amounts of uranium and thorium, which generate α-particles and are known to cause soft errors in sensitive circuitry, like dynamic random access memory (DRAM) cells.

To complete the molding compound formulation, small amounts of pigments, coupling agents, mold release agents, reaction accelerators, antioxidants, water getters, plasticizers, and flame retardants are all added. Coupling agents increase resin adhesion to the fillers, the chip, and the lead frame. Mold release agents do just that: help free the molded part from the mold chase. Flame retardants are a necessary requirement for the plastic package to meet the industry flammability standard of Underwriters Laboratories' (UL) standard 94 V-0, and until very recently, the standard was met by the use of brominated epoxy and antimony trioxide.

The properties of a molding compound are a balance between its moldability in a high-volume automated manufacturing environment and its relationship to the overall package's performance and reliability. The coefficient of thermal expansion is considered a good marker that correlates to the projected reliability, as it is a marker of the mechanical quality of the package and its ability to withstand thermal stresses. Flexural modulus of the compound is next in importance when it comes to reliability, as it indicates the amount of "give" the compound has in the presence of mechanical or thermal stress.

Table 6.2 Influence of Molding Compound Ingredient on Physical Properties

	Ingredients					
Property	Filler (>70%)	Epoxy (~10%) and Hardener (~7%)	Elastomer (<5%)	Catalysts	Flame Retardants and Scavengers	Waxes and Oils
Viscosity (rheology)	– – –	+++		– – –		
Cure rate (productivity)		+++		+++		
Mold cleanliness			0			00
Mold release	– – –	– –			+++	
Stress in device	+		+++			
Glass transition temperature (T_g)		0				
Strength	++	0	–			
Moisture absorption	+++	– – –	–	0		0
Thermal conductivity	+++					
Combustability	+++	–			+++	
Electrical reliability	+	0	–	–	0	0

Notes: +, positive influence; +++, strongest positive influence; –, unfavorable; 0, either favorable or unfavorable. This table provides a quick overview of the composition of typical epoxy molding compounds (EMCs) by major components and the effect that each of these components has on the key performance properties (molding and cured properties).

Source: Adapted from Richard C. Benson, Dawnielle Farrar, and Joseph A. Miragliotta, *Johns Hopkins APL Technical Digest*, 28(1), 58–67, 2008.

In summary, Table 6.2 shows how much effect each ingredient in a molding compound has on the overall material performance and behavior. As noted in the table, several key properties of the cured compound are attributable to the filler particles. The selection of filler type by material, size, and shape will control end parameters such as thermal expansion, moisture absorption, thermal conductivity, and strength.

6.1.4 *Newer formulations*

As already mentioned, epoxy cresol novolac was the backbone of most molding compounds until the 1990s. Then, with the use of larger and larger surface-mount packages and the advent of ball grid arrays and all of their attendant issues, new chemistries and formulations were required to meet their manufacturing and reliability needs.

6.1.4.1 *Biphenyl*

Biphenyl resins turned out to be the successful approach for reducing moisture uptake in compounds and increased resistance to popcorning (see Section 6.3.9). The reason was due to the nature of biphenyl resins that they could be loaded up with silica filler, to nearly 90% by weight. That way, there was hardly any organic material available to absorb moisture. The chemical structure is shown in Figure 6.1.

The possible disadvantage with biphenyl resins is their low glass transition temperature, typically around 125°C. It was feared that a given package using a biphenyl-based molding compound would most likely be subject to a coefficient of thermal expansion (commonly denoted as α_2) above the glass transition temperature during subsequent thermal processing and reliability testing. α_2 nearly always imposes a greater level of mechanical stress on a package. However, it turned out in numerous evaluations and field use that such fears about biphenyl molding compounds were unfounded, and these molding compounds were—and still are—used commercially.

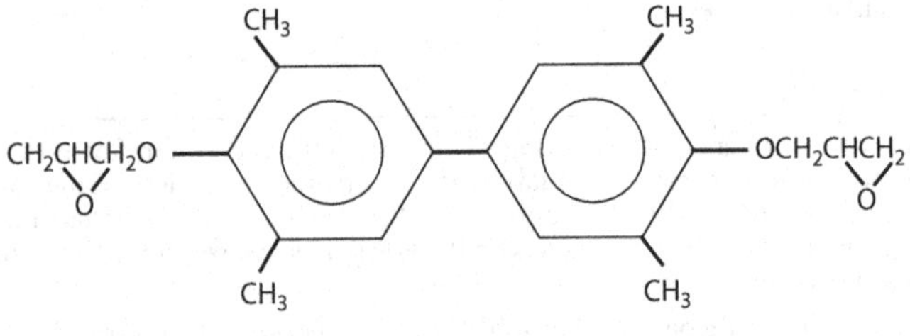

Figure 6.1 Chemical structure of biphenyl resin.

6.1.4.2 Multifunctional

Initially utilized to help resolve the *popcorn* package cracking issue, multifunctional resins have a high glass transition temperature (around 220°C) due to high cross-linking density. The resin structure is shown in Figure 6.2. The cross-linking density was thought to increase high temperature strength but turns out it also increases the resin's ability to absorb moisture. Therefore, multifunctional-based molding compounds turned out to be unsuccessful in combating the popcorn cracking issue.

However, it turns out that multifunctional compounds reduced the substrate package warpage issue, apparently because their high glass transition temperature meant the molding compound underwent less distortion than formulations with lower transition temperatures. Multifunctional-based molding compounds remain in commercial use for many types of substrate-based semiconductor packages.

6.1.4.3 Aromatic resins

With the requirement for environmentally friendly molding compounds (see Section 6.1.7.1), the resin backbone for molding compounds needed to change again. Examples of alternate resin structures that do not require bromine or antimony as flame retardants are shown in Figure 6.3. Many

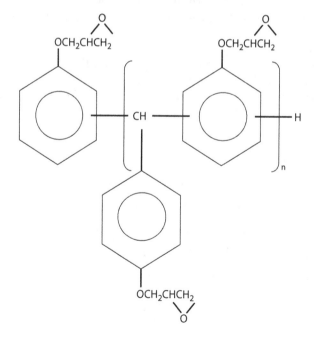

Figure 6.2 Chemical structure of multifunctional resin.

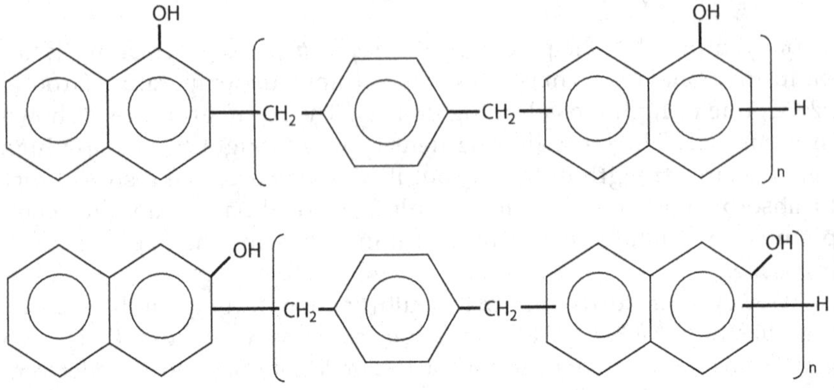

Figure 6.3 Examples of chemical structures for aromatic resins.

of these new organic materials are proprietary, and it is difficult to find much public information about their behavior.

6.1.5   Technology challenges

As packaging and manufacturing technology and processes have changed over the years, so have the formulations and behaviors of molding compounds to meet the needs of continued high-volume production.

6.1.5.1   Moldability

Transfer molding is the means of encapsulating the chip. Improving the moldability of a molding compound means controlling the viscosity and velocity of the melted compound before it sets up and hardens. Controlling

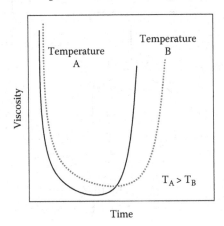

Figure 6.4 Viscosity curve versus time and under the influence of temperature.

these two variables is made more difficult due to the fact that molding compounds are non-Newtonian fluids; therefore, the fluid mechanics going on inside a mold chase are very complex. Figure 6.4 illustrates the behavior of viscosity over time as the molding compound undergoes heat and pressure during the transfer molding process.

The two factors must also be balanced between the need for high through-put while not sending the liquid compound through the mold with too much force and thus damaging the chip surface, or deforming the lead-frame lead tips or sweeping away the bonding wires.

Over time, the designs of mold chases have changed, going from a single, large-volume transfer pot into multiple cavities and runners to multiple, small-volume pot designs with short runners and few or two cavities to fill. Those changes also required tweaks to formulations.

6.1.5.2 Glass transition temperature

The glass transition temperature (T_g) represents the softening point of an adhesive. To measure that transition point, a differential scanning calorimetry (DSC) test system shows the peak exothermic reaction temperature for a given polymer system. Further information on DSC is given in Section 6.2 and in Appendix B.

6.1.5.3 Flexural modulus

The flexure test is a measure of the behavior of materials under simple beam loading or bending. Maximum stress and strain over increasing loads are measured and plotted in a stress–strain diagram. Flexural strength is defined as the maximum stress endured in the outermost fiber of the material, which is calculated at the test specimen surface on the convex or tension side. Flexural modulus comes from calculating the slope of

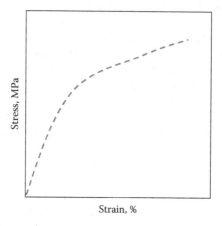

Figure 6.5 Stress–strain curve.

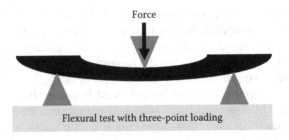

Figure 6.6 Flexural stress test using three-point loading.

the stress versus the deflection curve. Figure 6.5 shows the stress–strain curve from flexure testing.

Flexure testing is typically done on relatively flexible materials, like polymers, wood, and composites—which would include molding compounds. Test methods include three-point flex and four-point flex. The three-point flex example is shown in Figure 6.6, as this test method is most commonly used for polymers. In a three-point test, the area of uniform stress is rather small and concentrated under the center loading point.

6.1.5.4 Coefficient of thermal expansion

As stated earlier, epoxy alone has à very high coefficient of thermal expansion, around $80 \times 10^{-6}/°C$, compared to the silicon chip or copper lead frame or organic substrate. Filling the epoxy with silica filler brings down the thermal expansion coefficient into the teens, which would match the lead frame or substrate but would still be high compared to silicon. Using biphenyl or similar epoxy types allows more filler to be added and brings the coefficient down to the high single digits but still not as low as silicon, which is about $3 \times 10^{-6}/°C$.

Also, the glass transition temperature marks the point where the thermal coefficient increases significantly. Excursions into elevated temperatures may result in higher-than-anticipated stresses on the package and the device. The change of thermal expansion with temperature is illustrated in Figure 6.7.

In short, a molding compound's coefficient of thermal expansion is a tricky balancing act with the other components that make up a semiconductor package and the chip.

6.1.5.5 Stress index

This semiquantitative measure is generally used as an indicator for molding compound and the level of stress induced by the nature of the material. The stress index is defined as "E times CTE," or flexural modulus multiplied by coefficient of thermal expansion.

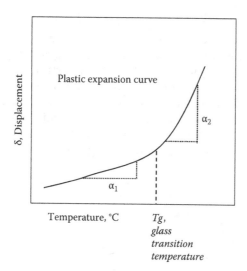

Figure 6.7 Change in coefficient of thermal expansion with temperature.

Stress indices are generally useful as a comparison tool, to rank various kinds of molding compounds for the levels of stress each might induce on a package and the die within. Lowering the stress in a molding compound involves meeting two opposing demands—lowering the coefficient of thermal expansion and lowering the compound's modulus. To lower the thermal expansion coefficient is to increase filler concentrations in the molding compound, which has the unfortunate effect of raising the flexural modulus. Lowering the flexural modulus is not so simple, but that usually involves altering the resin and hardener chemistry, making for more flexible polymer chains. However, those chemical changes are usually not enough to offset the high levels of filler required to lower thermal expansion.

6.1.6 Failure modes associated with molding compounds

Along with the aforementioned issues with new flame retardant materials, the most prominent issue associated with molding compound is solder reflow package cracking, otherwise known as *popcorning*.

6.1.6.1 Package cracking during solder reflow

As discussed in Chapter 3, when large and thin (3-mm thick and below) surface-mount packages (such as quad flat packs and plastic leaded chip carriers, and later, plastic ball grid arrays) became prevalent in the late 1980s, the phenomenon named *popcorning* became well known. Popcorning occurs when a moisture-filled plastic package undergoes the solder reflow process in an in-line oven. The moisture often pools in areas of weakened adhesion, which eventually causes delamination. The package is subjected to rapid heating to temperatures above the boiling point of water. Therefore, the moisture inside the package turns to steam and exerts pressure to escape. The molding compound typically cannot withstand the force and commonly fails through the backside, resulting in an external crack. The sound of the package failing is said to sound like popcorn popping, and hence the term *popcorning*. The sequence of events was shown in Figure 5.4 and is repeated again

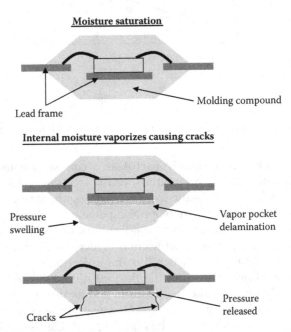

Figure 6.8 Sequence of events during package cracking, or *popcorning*, event.

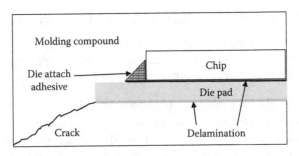

Figure 6.9 Package material interfaces prone to delamination and subsequent package cracking (not to scale).

in Figure 6.8, and Figure 6.9 (once again from Figure 5.3) shows the end result of popcorning.

A standard—the IPC-SM-786A—was created by the IPC® (formerly the Institute for Interconnecting and Packaging Electronic Circuits) to control moisture levels in semiconductor packages prior to solder reflow. The specification, now superseded by J-STD-020 and J-STD-032, specified storage conditions and the need for dry-packing or even baking before reflow for very sensitive packages.

Still, there was great interest into the 1990s to discover chemistries that could make molding compounds that were less moisture sensitive. Research in this area led to the development and use of biphenyl-based molding compounds.

6.1.6.2 Substrate postmold warpage

With the advent of plastic ball grid arrays and other substrate-based plastic semiconductor packaging, another issue arose, that of substrate warpage, and not subtle or slight concave or convex curvature, but readily obvious to the naked eye, as shown in Figure 6.10. What can be more alarming is that the type of curvature will change depending on the temperature and thermal processing. In all, such a high level of warpage and cycling between negative and positive modes naturally exerts unwanted stresses on the package, affecting performance and reliability.

Causes are multiple, but they include the fact that organic substrates are much less rigid than metallic lead frames and that molding compound is applied to only one side of the substrate. The thermal and mechanical

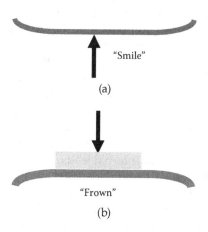

Figure 6.10 (a) Concave curvature ("smile") of substrate and package (not to scale). (b) Convex curvature ("Frown") of substrate and package (not to scale).

properties of the molding compound need to be in better balance with the properties of the other packaging materials.

6.1.7   Future developments

Just as package technology undergoes continual changes and new types of packages are developed, the properties and characteristics of the molding compounds used to encapsulate them evolve.

6.1.7.1   "Green" molding compounds and changes to flame retardant additives

The push for "green" or environmentally friendly materials received a big impetus from the effect of environmental regulation on electronic systems, specifically with the RoHS (Restriction of Hazardous Substances) directive adopted by the European Union in February 2003, which came into force in July 2006. This directive directly affected molding compound composition due to its regulation about the levels of bromine allowed.

For some background, to meet Underwriters Laboratories' flammability standard V-0, molding compounds traditionally employed brominated epoxy resins and antimony oxides as flame retardants. But with the amount of organic compounds with bromine restricted, molding compound formulators had to come up with alternate flame retardant solutions.

The road to create molding compounds that meet these new environmental regulations has not been without its bumps. Sumitomo Bakelite, a leading supplier of molding compound, first used coated inorganic red phosphorus particles as substitute flame retardant material in its EME-U family of "green" molding compounds during the late 1990s (the coating of aluminum hydroxide and phenol resin was necessary to prevent the diffusion of red phosphorus ions throughout the compound). Otherwise the new compounds met all the mechanical, physical, and reliability criteria necessary for a production-worthy product.

However, after less than a year of field use (in some cases, only a few weeks), devices packaged with the new formulation began to fail. The most common failure mode identified was resistive short and current leakage between adjacent internal leads. Often, the failures appeared intermittently, and seemed to be dependent on humidity, voltage, and temperature. Another failure mode seen was increased resistance or an open circuit due to wire bond failures.

Extensive failure analysis revealed the phosphoric acid caused copper and silver migration and filament growth, causing shorts between the leads. As it turns out, the coating around the red phosphorus was either too thin or not sturdy enough to withstand all the process and manufacturing steps. The aluminum hydroxide coating could degrade in the presence of moisture and oxygen during elevated temperatures, thus converting into alumina and water as shown in Equation (6.1):

$$2Al(OH)_3 \rightarrow 3H_2O + Al_2O_3 \tag{6.1}$$

The resulting phosphoric acid now acts like an electrolyte, allowing for copper and silver migration, especially if adjacent leads are electrically biased. The process is illustrated in Figure 6.11. It was also found that an excess of phosphoric ions probably contributed to accelerated ball bond pad intermetallics and corrosion, leading to the wire bond failures.

In turn, the release of phosphorus ions (PO_4) is also increased in the same presence of elevated temperatures and increased moisture. Such elevated temperatures occur normally during manufacturing, like at the transfer molding step as well as at solder reflow. With the increased oxygen and water present, the phosphorus ions are converted into acids, as seen in Equation (6.2):

$$4P + 5O_2 + 6H_2O \rightarrow 4H_3PO_4 \tag{6.2}$$

In the end, red phosphorus proved not to be the solution to the "green" flame retardant issue, as it proved too reactive in the typical environment seen by plastic semiconductor packages. The coatings solution to prevent reactivity ended up not being effective enough.

Although red phosphorus proved disastrous, current "green" molding compound formulations use other proprietary solutions, such as transitional metal oxides.

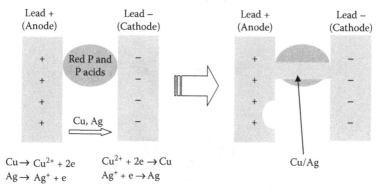

$$Cu \rightarrow Cu^{2+} + 2e \qquad Cu^{2+} + 2e \rightarrow Cu$$
$$Ag \rightarrow Ag^+ + e \qquad Ag^+ + e \rightarrow Ag$$

Figure 6.11 Red phosphorous and phosphoric acid acting as a catalyst for copper and silver migration between lead frame leads (not to scale).

6.1.7.2   Molded underfill

This product aims to combine the manufacturing and physical properties of a compound with that of flip-chip underfills. Although the two products have many similarities in chemistry and composition, their purposes, and therefore behaviors, are different. It has proven tricky to match the two primary purposes—fully filling the gap between chip and substrate with good moldability in high-volume production.

More details about the progress in molded underfills will be discussed in Section 6.3.

6.1.7.3   High-density packaging

High-density packaging refers to packages with multiple chips, whether side-by-side or stacked on top of one another. For complex structures involving multiple chips and multiple levels of wire bonding, there is the risk of excessive wire sweep and subsequent yield loss. Some level of wire sweep could be mitigated by optimizing the layout of the bonding wires, but careful selection of a molding compound formulation is also necessary. In addition to any changes to molding compound to reduce wire sweep even further, usually by going to very low viscosity formulations, there may be changes necessary to the mechanical aspect of molding, such as the mold chase design or even moving away from traditional transfer molding methods altogether.

One such change to the mold chase design is going from the conventional bottom-gate-in-the-corner design to a top-gate or center-gate design, as shown in Figure 6.12. A radial mold flow from a top gate could minimize

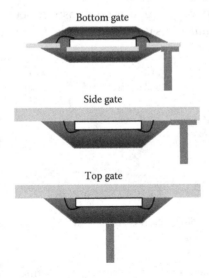

Figure 6.12 Comparison of mold gate placement.

wire sweep and reduce filler separation, which can occur when the fine pitch bonding wires filter out fillers as the molding compound moves between them. This becomes more and more of a problem as the wire lengthens.

Another solution may be dispensing an encapsulant over the wires prior to the molding step, in order to "lock down" the wires in place. However, that adds a few more process steps to the manufacturing flow, which tends to be undesirable to through-put and cost but could be justified with higher yield.

Some alternate methods to transfer molding might be liquid molding, compression molding, and transfer molding with vacuum assist. These alternate processes remain in the developmental stage.

6.1.7.4   *Compatibility with copper wire bonding*

With the high price of gold (hovering near $1,000 per ounce) becoming a norm, back-end operations for semiconductor packaging are looking to substitute copper wire for the wire bonding step, as a cost-control measure. The advantages and disadvantages of copper wire versus gold are examined in detail in Section 6.2.

Here, the discussion focuses on how the change in bonding wire changes the performance requirements of the molding compound. For instance, recent studies point toward more careful control of chloride levels to ensure a reliable package when exposed to moisture. (There are some indications that perhaps pH levels also need to be controlled, but this has not yet been confirmed.) Copper behaves differently when bonding onto aluminum bond pads. In general, copper is more reactive than gold and corrodes more easily. Apparently the copper–aluminum bond interface is very susceptible to chloride corrosion, especially in the presence of electrical bias.

Even without the presence of moisture, the interaction between molding compound and copper wire bonds is different than that with gold wire. With only heat as the acceleration factor, copper wire bonds still showed greater corrosion susceptibility than bonds made with gold wire.

Given the data to date, molding compound selection for copper wire bonding points toward "green" molding compounds, because they have very low ionic levels to begin with to meet RoHS regulations.

6.2   Die attach adhesives

6.2.1   Objectives

- Describe a die attach adhesive.
- Convey its purpose in semiconductor packaging.
- Discuss some of the unique reliability and performance issues associated with die attach adhesive.

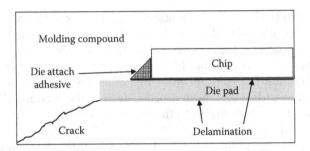

Figure 6.13 Package material interfaces prone to delamination and subsequent package cracking (not to scale).

- Illustrate some of the ongoing research and development of this material.

6.2.2   Introduction

The history and use of adhesives to attach chips to lead frames and substrates in plastic semiconductor packages is described here. Figure 6.13 (which is the same as Figure 5.3) illustrates how die attach adhesive is used in semiconductor packaging to create a stable platform for the next manufacturing step, which is electrically connecting said chip through wire bonding from the die pads to the lead frame or substrate leads.

6.2.3   Background

The die attach process step is affixing the silicon chip (or chips) to a lead frame or substrate with an adhesive or metallic solder in the form of a paste, a film, or solder wire or preforms. The chip is attached from its backside to the metal surface of the lead frame or bonding area on a substrate. The connecting material usually offers thermal and electrical conductivity, or sometimes thermal conductivity alone.

For the purposes of brevity, the discussion in this chapter will be limited to organic-based adhesives used as a die attach. Like molding compounds, die attach adhesives using organic materials appeared in conjunction with the emergence of plastic semiconductor packaging in the 1970s. Prior to plastic packages, semiconductors were encased in hermetic packages, either of ceramic or metal construction, and the die attach material used was formed of solder, typically a rigid gold-silicon eutectic or a softer solder compound, or silver-filled glass. Moving to a polymer-based adhesive offered such advantages as lower induced stress, lower processing temperatures, and lower material cost. For those reasons, in nonmilitary or aerospace applications, even in the assembly of ceramic packages, the preference would be to use a polymeric die attach, unless other considerations precluded that choice.

And like molding compounds, the base material initially chosen as the adhesive was an epoxy resin. However, early adhesive formulations used an epoxy with aliphatic amines as the curing agent. Although aliphatic amines were highly reactive, they unfortunately did not react fully with the epoxy resin. These solvents were prone to high levels of outgassing, leaving voids in the adhesive layer, which then led to poor bond strength, and contaminant material deposited on the die and package surfaces, which then led to delamination between interfaces. Also, the unreacted amines would later react in the presence of moisture to free up chlorine ions (Cl^-), which then caused corrosion failures on the die or within the package.

Given these corrosion problems, people turned next to preimidized polyimides as an adhesive solution, given their low levels of hydrolizable chlorine ions. This type of polyimide remained popular as a die attach adhesive starting in the late 1970s through the late 1980s. Though the risk of corrosion was reduced, polyimide in the form used in die attach adhesives—polyamic acid in N-methypyrolidinone (NMP) solvent—had several disadvantages. One was a long two-step cure profile that took anywhere between a half-hour to 2 hours, reducing manufacturing throughput. Another was the elevated temperatures (250 to 350°C) necessary to fully convert the polyamic acid to polyimide. Because of the evolution of solvent and water during the cure, polyimides would see a large weight loss and shrinkage, resulting in a high elastic modulus, which placed undue mechanical stress upon large-sized chips. Finally, cured polyimides tended toward high moisture absorption, which reintroduced the risk of corrosion again if ionic contamination from other sources was present.

By the mid-1980s, new epoxy resin systems were available, with the amount of chloride reduced and overall ionic levels down to 10 ppm or less. Along with the addition of silver flakes for thermal and electrical conductivity requirements, silver-filled epoxies became the predominant die attach adhesive solution from the late 1980s onward.

During the 1990s, research and development on die attach adhesives focused more on manufacturing issues, such as better rheology to reduce incidents of tailing and stray material deposited in undesirable areas seen during very high-speed dispensing during production. The other area of focus was reducing the cure time necessary from the typical 1 hour to a few minutes or even less than a minute. In the last case, the goal was known as *snap-cure*.

There were other incremental developments to improve various properties of epoxy adhesives, such as reducing moisture uptake and lowering the elastic modulus to make the die attach "low-stress" for larger-area chips.

6.2.4   Materials composition

To be specific, an epoxy die attach adhesive is typically made up of the following components: liquid epoxy resin, silver flakes or some kind of filler material, reactive epoxy diluent or solvent, catalyst and hardener, and other additives.

6.2.4.1   Liquid epoxy resin

The liquid epoxy resin is the base for the adhesive system. Its most important characteristic is "cleanliness," or lack of ionic contamination.

6.2.4.2   Silver flakes and other filler materials

For die attach adhesives, silver can be found in either powder or flake forms. Flakes have a much greater surface area, and they tend to overlap each other on contact, resulting in a high surface conductivity. However, flakes tend toward agglomeration, which can be treated by coating the flakes with fatty acids, but that will reduce conductivity slightly. Nonetheless, silver is a good conductor, even in its oxide form. Other metals' oxides, like copper and aluminum, have much lower conductivity. For compositions not requiring electrical conductivity, thermally conductive fillers are used, like alumina or other ceramics.

   The particle size and distribution of the silver flakes are key to the rheology and adhesion behavior of the die attach adhesive.

6.2.4.3   Reactive epoxy diluents and solvents

The intent of a reactive diluent is to adjust the adhesive's rheology while curing, but minimize the amount of outgassing that would typically be seen if a solvent was used in the adhesive system. Ideally, the reactive diluent would react with the polymer matrix before the gelation stage is completed and not release materials that would redeposit on the die and package surfaces and create contamination. Of course in the real world, there is always some level of outgassing, but the goal is a minimal level.

6.2.4.4   Catalysts and hardeners

Various factors go into selecting a catalyst and hardener: length of pot life (usage time out on the factory floor), shelf/storage life, level of adhesion, and level of induced stress. For instance, a lesser amount or a slower-acting catalyst likely mean a longer pot life but also longer time required for cure.

6.2.4.5   Other additives

Additives are introduced into the adhesive formulations to reduce stress levels, increase adhesion, reduce resin bleed, and tie up any residual ions.

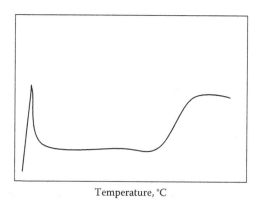

Temperature, °C

Figure 6.14 Example of a differential scanning calorimetry curve (not to scale).

These components are usually proprietary and play an important role in producing a high-performance die attach adhesive.

6.2.5 Materials analysis

For military applications, a die attach adhesive must conform to Mil-Std-883, test method 5011, which also covers polymeric adhesive specification and requirements.

6.2.5.1 Glass transition temperature

As for molding compounds, the glass transition temperature (T_g) represents the softening point of an adhesive. To measure that transition point, a DSC test system shows the peak exothermic reaction temperature for a given polymer system, like that of a die attach adhesive.

Figure 6.14 shows an example of a DSC curve to determine the T_g for a polyimide die attach adhesive after a two-step cure process.

6.2.5.2 Coefficient of thermal expansion

As stated earlier, epoxy alone has a very high coefficient of thermal expansion, around 80×10^{-6}/°C, compared to the silicon chip or copper lead frame or organic substrate. Die attach adhesives are not as heavily filled as molding compounds, and often the filler is conductive silver flakes, to meet the need for electrical and thermal conductivity. Generally, a silver-filled epoxy adhesive will have a thermal expansion coefficient of around 20 to 30×10^{-6}/°C.

6.2.5.3 Thixotropic index

Thixotropic materials have a variable viscosity. The viscosity typically decreases with time under high shear rates. This property is reversible.

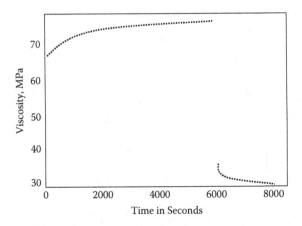

Figure 6.15 Example of thixotropic behavior using a 5% bentonite solution (not to scale).

An example of thixotropic behavior is shown in Figure 6.15, which used a bentonite solution as its test medium under various shear rates. First, the bentonite solution is presheared at a shear rate of 100 s⁻¹ for 5000 seconds before experiencing a shear rate of 25 s⁻¹ for 6000 seconds before returning to a shear rate of 100 s⁻¹.

The accepted explanation for thixotropic behavior is that the viscosity of such materials (usually consisting of tiny particles suspended in a fluid, and precisely describes die attach adhesives) is such that said particles interconnect or clump together and resist being rearranged. When under shear, the particle structures break down in time, and therefore, the viscosity drops off. If the material is then left to rest unstressed, the particles rearrange themselves into structures, and the viscosity subsequently increases.

6.2.5.4 Ionic purity

As stated earlier, eliminating as much of the ionic contamination by alkali metals (sodium, potassium, and the like), chloride ions, and other impurities present in the organic resin is imperative to the prolonged reliability of the package and the device. Ionics and moisture can become a deadly combination, causing corrosion and electrical shorts and resulting in premature failure.

Measuring ionic impurity levels can be done with ion chromatography. An example of an ion chromatography reading, or chromatogram, is shown in Figure 6.16.

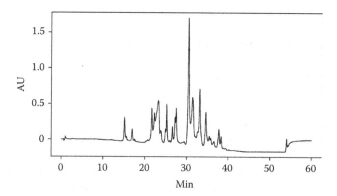

Figure 6.16 Example of ion chromatogram. (From Wikimedia Commons, attributed to user Lukke.)

6.2.6   Reliability and performance

Much like molding compounds, the performance of die attach adhesives has a lot to do with adhesion strength, low moisture uptake, and coefficient of thermal expansion compatibility. In addition, there are several other factors for consideration, already touched upon in Section 6.2.3: outgassing and resin bleed.

6.2.6.1   Outgassing

Outgassing refers to the release of aromatic compounds during cure. Initially in a vapor state, the residues eventually settle onto a surface. Unfortunately, sometimes that surface is the chip passivation, or the lead frame, or the substrate. The residue will reduce adhesion, which will eventually become delamination.

To prevent outgassing residues from affecting the integrated circuit and package, elevated airflow is employed during the curing step, to "blow away" any contaminants.

6.2.6.2   Resin bleed

Also referred to as *bleedout*, resin bleed is the separation of the resin vehicle from the polymer adhesive. The problem comes from the excess resin coating the die paddle or substrate surface not covered by the chip and reducing adhesion with the overmold. The other issue is resin coating any wire bonding pads, which then would then adversely affect ball bond-to-pad adhesion.

Paste adhesives are most prone to resin bleed, while film adhesives are less so.

6.2.7 Future developments

Much like molding compounds, the properties and characteristics of die attach adhesives must evolve as package technology undergoes changes and new types of packages are developed.

6.2.7.1 Three-dimensional (3D) packaging

With the advent of stacked-die and three-dimensional (3D) packaging, the need for controlled minimal thickness and no bleeding has resulted in the popularity of die attach film rather than adhesive paste in such packages. Figure 6.17 (already shown as Figure 2.8) shows an example of a 3D package structure and illustrates why a film die attach adhesive would be desirable, as there would be little or no bleedout onto the bond pads.

The important factors that die attach adhesive aimed at such complex structures is maintaining material properties and their interactions with other parts of the system. Properties such as adhesion, shrinkage, cure temperature, modulus, moisture absorption, and coefficients of thermal expansion must all be carefully controlled. Physical and chemical stability through multiple heat cycles during assembly is also extremely important.

New material formulations have been developed to address the unique requirements for multiple die stacks with wire bond interconnects. Traditional epoxy die attach materials tended to have a high modulus, which resulted in high stresses within the stack and, therefore, cracked die. Traditional fillers used in nonconductive die attach adhesives tended to be very abrasive which left damaged passivation and led to electrical failures. Therefore, the new formulations needed to be low stress and nonabrasive.

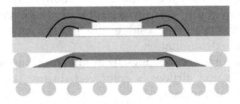

Figure 6.17 Package-on-package (not to scale).

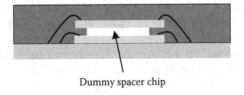

Dummy spacer chip

Figure 6.18 Stacked-die package with dummy spacer chip (not to scale).

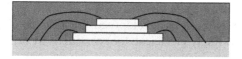

Figure 6.19 Stacked-die package with a pyramid-type chip stack (not to scale).

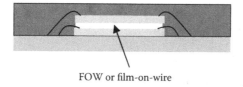

FOW or film-on-wire

Figure 6.20 Stacked-die package using film-on-wire (not to scale).

An additional factor to be considered is when stacking chips of similar size (Figure 6.18) rather than a pyramid-type stack (Figure 6.19). Often, a dummy space chip is employed to leave enough clearance for the wire bonds. More recently, die attach adhesives dubbed *flow-over-wire* (FOW) have been developed to eliminate the need for spacer die—and the additional manufacturing steps to attach them—as well as offer enough rigidity during the wire bonding steps but without inducing stress or surface damage to the chips. And the FOW material encapsulates the wire bonds and protects them.

Creating a FOW film material may further consolidate manufacturing steps by eliminating the adhesive dispense and cure process steps. An example of how a FOW stack might look is shown in Figure 6.20.

6.2.7.2 *Lead-free and restriction of hazardous substances (RoHS)*
The introduction of lead-free solders and other "green" initiatives means higher temperatures during assembly processing. Thus, die attach adhesives, like the rest of the material set, must adapt to the elevated thermal environment. Also, the adhesives must conform to environmental regulations, and the amounts of hazardous (lead-free, halogen-free) materials may be limited.

6.2.7.3 *Compatibility with copper wire bonding*
Much like molding compound, improved reliability is observed when using a "green"-designated die attach material. As mentioned in the last chapter, recent studies point toward more careful control of chloride levels to ensure a reliable package when exposed to moisture. "Green" packaging materials, including those of die attach adhesives, help to ensure very low free ionic content within the package.

6.2.7.4   Other developments

Other areas of research and development include developing low-stress, high thermal (and likely electrical, too) conductivity adhesives to counter hot spots within complex system-in-package designs. Also needed is an adhesive material that is compatible with low-κ dielectric materials, especially when such chips are stacked upon each other.

6.3   Underfill materials

6.3.1   Objectives

- Describe an underfill material.
- Convey its purpose and importance in flip-chip packaging.
- Discuss criteria important in evaluating reliability and field performance.
- Illustrate some ongoing research and development of this material.

6.3.2   Introduction

This section discusses the need and use of underfill material in flip-chip technology, especially with organic substrates. As the name implies, an *underfill* fills in the gap between the flipped chip and the substrate the chip is connected with. Underfills lend necessary mechanical strength and environmental protection to the flip-chip assembly.

6.3.3   What is underfill?

Underfills, especially those used with capillary dispensing, are generally epoxy-based materials filled with fused silica at up to 70% by weight, much like molding compounds. Of course, the composition and behavior differ from molding compound, mostly due to the application method. As implied, underfills typically encapsulate the chip–substrate interface via capillary flow. They were derived by chip-on-board encapsulants, which have existed since the beginning of the semiconductor industry, when liquid epoxies were used to pot transistors.

6.3.4   The purpose of underfill

Underfill encapsulants are just as described: the encapsulant fills the space and tightly couples the flipped chip and organic substrate, in order to compensate for the coefficient of thermal expansion mismatch between the silicon (at about $3 \times 10^{-6}/°C$) and the organic printed circuit board (PCB), which is closer to $20 \times 10^{-6}/°C$. Without the underfill, the thermal expansion mismatch will cause undue mechanical stresses on the solder joint connections and would lead to premature failures. The underfill redistributes the stress profile by transferring the shear stresses experienced by the solder joints into bending stresses withstood by the entire package structure. Without the use of underfills, flip chip on organic substrates would likely be impossible, except perhaps for very small chips, and its use never would have proliferated.

In the case of ceramic substrates, such as those used in IBM's C4 (Controlled Collapse Chip Connection) process, underfill is generally not necessary, at least for mechanical stresses, because the coefficients of thermal expansion of silicon and alumina (about $7 \times 10^{-6}/°C$) are relatively similar.

Underfills fulfill secondary purposes, such as protecting the chip and interconnects from other environment factors, such as corrosion caused by moisture ingress. Underfills can also help absorb alpha-particle emissions, which can cause soft errors in the chip's circuitry, especially memory devices. The lead used in solder can be a source of alpha particles.

In the present, underfill materials must often fulfill competing requirements. They must behave well and consistently during the manufacturing process, often needed to flow rapidly through ever-narrowing gaps and openings without causing voids. Once in place and cured, the underfill must protect the flip-chip interconnects and the chip's active circuitry from mechanical damage. It also cannot induce any damage. The underfill must remain durable during continued and repeated cycles of temperature, moisture, and environmental exposure.

There is one more factor to consider in any interaction between solder flux used in the interconnection and the underfill. Fluxes used for lead-free solders tend to be more active and may adversely react with the underfill. If a no-clean flux is used, then it is guaranteed that there will be some residue present, and its interaction with the underfill material should be understood.

6.3.5   The (standard) underfill process

An overview of the underfill process by capillary action:

- The flip chip is conveyed in-line or hand-loaded into the dispensing equipment.
- The die is then heated to temperature or is heated during dispensing to provide good underfill flow.
- The flip chip is located either mechanically by fixture or by an automated vision alignment system.
- The fluid is dispensed on one or more sides of the flip chip, sometimes in multiple dispense passes, as shown in Figure 6.21.
- The underfill fluid flows via capillary action under the flip chip.

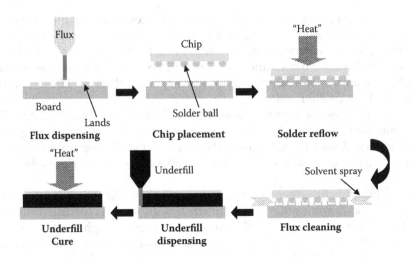

Figure 6.21 Overview of the underfill process using capillary action.

- Depending on the pattern chosen, a fillet pass may be required to provide an even fillet around the perimeter.
- The underfill is cured in a reflow or microwave oven at the underfill manufacturer's recommended temperature.

The main complaint with the use of underfills is that they add additional processing steps and equipment to the flip-chip assembly process, again as shown in Figure 6.21. Admittedly, waiting for an underfill to fill the gap by capillary action is not an instantaneous process.

However, underfills have proven their necessity by improving the reliability of flip-chip packages. Capillary underfills are a proven materials set, and there has been a long history in optimizing the manufacturing process. Therefore, many attempts have been made to speed up the process or try other approaches to reduce the number of steps and increase overall production through-put.

6.3.6 Underfill properties

Various materials properties must be considered in making a "good" underfill, in respect to manufacturing and reliability considerations. For one, a long pot life is important for high-volume production, as the material properties should remain stable for at least one production shift or 8 hours.

Filler particle size is another important factor, as a balance is needed between filler weight percentage to lower the thermal expansion coefficient and ensure good capillary flow during manufacturing. Also, filler particles cannot be too large, or there might be flow blockage and, therefore, voids. Preventing filler separation is a subset of filler content, as it is important to have consistent filler distribution to maintain consistent thermal expansion properties.

After curing, other material properties are considered important for reliability. In addition to the aforementioned coefficient of thermal expansion, they include elastic modulus, glass transition temperature, tensile strength, and moisture resistance.

Finally, the adhesive strength of the cured underfill to all of the various interfaces—solder, passivation, solder mask—is critical to maintain package integrity. Cracks and delamination risk premature package failure, either through contamination, moisture degradation, or stress concentrations on solder joint. Many factors affect adhesion levels—underfill formulation and wetting characteristics, cleanliness and condition of the surfaces prior to underfill application, type of solder mask and passivation, plasma treatments to enhance adhesion, and even manipulating cure profiles to maximize cross linking in the underfill.

6.3.6.1   Glass transition temperature

As for molding compounds and die attach adhesives, the glass transition temperature (T_g) represents the softening point of an adhesive. To measure that transition point, a DSC test system shows the peak exothermic reaction temperature for a given polymer system, like that of a die attach adhesive (Section 6.2) and a molding compound (Section 6.1).

6.3.6.2   Coefficient of thermal expansion

As stated earlier, epoxy alone has a very high coefficient of thermal expansion, around $80 \times 10^{-6}/°C$, compared to the silicon chip or copper lead frame or organic substrate. Underfills do not contain as much filler as molding compounds in order to flow more easily out of a needle, so, consequently, the thermal expansion coefficients for underfills are much higher than those for molding compounds, and are closer to die attach adhesives.

6.3.7   Alternate underfill processes

These approaches have been investigated and reported in books and literature for many years, but it is unclear whether they have taken hold in high-volume production, at least judging from the most recent datasheets from leading semiconductor assembly and test service suppliers, the exception being molded underfill usage publicly announced by Amkor Technology.

6.3.7.1   "No-flow" underfill

One well-documented approach is through the use of *no-flow* underfill. By no-flow, an underfill material is applied to the chip before it is joined to

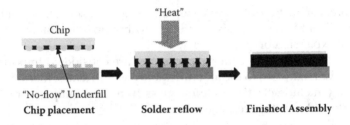

Figure 6.22 "No-flow" underfill process steps.

the substrate. The idea here is the underfill would spread and fill the gap, and then be cured at the same time while the solder joints are reflowed. The addition of flux to the underfill compound could save another step and eliminate another item from the bill of materials. The concept is very attractive, as it would combine three steps—solder joint reflow, underfill dispersion, and underfill curing—into one, as illustrated in Figure 6.22.

However, there are several challenges to the approach, and these seem to have kept no-flow away from high-volume production, at least in conjunction with solder joints. One is that voids tend to get trapped during the rapid curing process if used with solder joints. Another potential issue is that, unlike capillary underfills, no-flow materials are unfilled products by necessity. Fillers in underfills can hinder contact between the solder balls and pads. However, without the silica fillers, the thermal expansion mismatch remains more pronounced, and temperature cycling reliability tends to be reduced.

The no-flow concept has more promise when the interconnection is achieved with conductive adhesives or gold stud bumps, because rapid cure is not necessary.

6.3.7.2   *Reworkable underfill*

Reworkable underfills address another shortcoming—thermosets like epoxies cannot be reused or even be easily detached when it might be necessary to replace or rework the solder joints. This might not be an issue with boards used in portable consumer electronics—per board cost may be low and not warrant the extra expense of rework versus recycling the board's materials—but salvaging a board that might cost several hundred dollars or more, such as in a server, makes reworking much more attractive.

Reworkable underfills are often thermoplastics, which soften upon reheating, allowing for easier chip removal from the board.

6.3.7.3   *Preapplied underfill*

Preapplied underfill refers to an underfill material applied to the finished wafer, before any back-end processing has started. The underfill

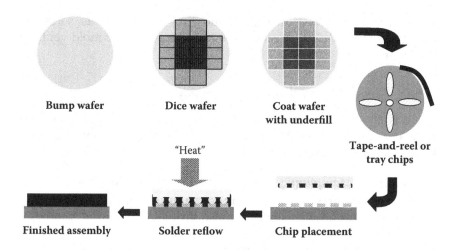

Figure 6.23 Process steps for preapplied underfill material.

application may even precede that of wafer bumping, though it more likely occurs afterward.

The concept is appealing, as it would eliminate several steps for back-end assembly, but there are several issues to consider (Figure 6.23). Applying the underfill material to the wafer would require very tight process control. Any underfill material applied prior to bumping must be able to withstand subsequent processing. Conversely, materials applied after bumping must not damage the bumps. Underfill stability is also a consideration, if the precoated wafers or chips will be stored for a period of time before assembly.

6.3.7.4   Molded underfill

Underfilling the die and then overmolding the entire structure to match a standard package outline adds several process steps and slows manufacturing through-put. As already mentioned in Section 6.1.7.2, Molded underfill, the concept of molded underfill offers a simplified process solution with high through-put along with the added advantages associated with the well-established transfer molding technology.

Another advantage of going with a molded underfill is eliminating the underfill fillet and any resin bleeding. This allows for more area available on the substrate to place decoupling capacitors and also for placing them closer to the chip, for improved electrical performance.

On the other hand, care must be taken when designing the mold chase for a molded underfill, to ensure good flow underneath the chip to prevent air pockets and voids from forming. Due to the nature of flip-chip geometries, there is a higher resistance to material flow and thus more risk of voids being formed. One solution to this problem is to use a vacuum-assist during the molding process. Other solutions might use the addition of vent holes in the substrate or mold vents in the chase.

6.3.8   Areas of research and development

There is always room for improvement in underfill behavior and performance, especially as the standoff height shrinks and bump density increases with the growing number of inputs and outputs (I/O).

6.3.8.1   Maintaining capillary flow as features sizes shrink

As die size gets larger while the features on the chip shrink, and the gap height between silicon and substrate also shrinks, maintaining good capillary flow becomes increasingly difficult. It takes longer to fill in general, affecting manufacturing through-put. The areas with bumps inside the gap will affect the underfill flow differently than those without. The mixing of different flow fronts across different parts of the chip can result in voiding or streaking (streaking being areas of slow or static flow and that could result in filler settling).

In this case, and the same with molded underfills needing to fill very narrow gap widths, filler technology may turn to nanoparticles as a solution. Nanoparticles may be able to resolve some of the issues associated with larger filler sizes, such as filler settling, clogged flow, and hindering solder wetting, though the tiny particles will probably bring their own set of issues, like agglomeration.

6.3.8.2   Compatibility with lead-free bump process steps, including for copper pillar bumps

The underfill material is designed for toughness as well as adhesion to various interfaces—solder resist on substrate, passivation on die, copper-bump, lead-free solder, and silicon—under lead-free, higher-temperature reflow conditions. These requirements mean careful choices must be made of the epoxy resin chemistries used in underfill material development.

Given many new chip and substrate surface chemistries will enter production in the near future, so will underfill chemistries need to change to meet manufacturing and reliability requirements.

6.3.9   Failure modes

Underfill materials are known to crack during temperature or power cycling, either during reliability testing or during actual field operation. The thermal expansion mismatch between the chip and underfill causes stress concentrations in certain locations, like at a die corner. As the crack propagates during continued stressing, the crack opens up the chip–underfill interface or another weakened section, causing either interconnect fatigue or an electrical failure in the chip's dielectric.

Other issues will arise if underfill flow is insufficient or creates voids underneath the chip, again becoming areas of stress concentrations or where moisture can pool and cause corrosion. And speaking of corrosion, incomplete flux cleaning could leave excessive corrosive elements within the package or cause delamination between interfaces.

6.4   Organic substrates

6.4.1   Objectives

- Describe what constitutes an organic substrate.
- Convey its purpose and importance in semiconductor packaging.
- Discuss the many different technologies and form factors available for substrates.
- Illustrate some of the future developments for substrates.

6.4.2   Introduction

Substrates or chip carriers have been around since the beginning of semiconductor packaging technology, generally in the form of ceramics or other hermetic materials. However, epoxy-impregnated glass fabric has been used for circuit boards for decades. Then in the late 1980s, Motorola and Citizen jointly developed the plastic ball grid array (PBGA). The widespread adoption of the PBGA in the 1990s led to the development of a variety of substrate-based packages, both using wire-bonded and flip-chip interconnects, and finally leading to stacked-die, or 3D, packages on substrates.

Therefore, the substrate carrier has become a very important part of semiconductor packaging, with the complexity and detail of a miniature circuit board. The material components that make up a substrate are also important, as they help determine the electrical, mechanical, and reliability performance of a given package.

6.4.3   Background

The construction of an organic substrate for semiconductor packaging has been covered in many technical books, journal articles, and industry-related publications. Organic laminates generally consist of a polymer resin—like FR-4 epoxy or bismaleide triazine (BT)—reinforced glass fiber fabric and conductive material, like a copper film coated with solder on the exposed surface, etched into lines and circuitry, and all protected by a coating of organic solder mask.

The coefficient of thermal expansion for an organic PBGA substrate is generally around $17 \times 10^{-6}/°C$. This results in a thermal mismatch with the silicon chip, but the molding compound or underfill encapsulant serves to bridge the difference and reduce mechanical stresses on the package.

The glass transition temperature for FR-4 is around 125°C, and it is about 175°C for BT substrates. The higher glass transition temperature for BT is why it became a dominant choice for package substrate material, as many processes, including solder reflow, occurred at temperatures of 175°C and above.

6.4.4 Ball grid arrays and chip scale packages

As the feature sizes on semiconductors dropped below the submicron level, the design pitches within packages also shrank, even as the I/O numbers increased in a smaller and smaller footprint. Traditional plastic packages utilizing metal lead frames and peripheral leads could not keep up with the shrinking die, and it was becoming very difficult to board mount quad flat packs with lead pitches below 0.5 mm. So, packaging turned to substrate-based packages with area arrays of solder balls on the bottom side, at pitches going well below 1 mm. Figure 6.24 illustrates a typical cross section for conventional wire-bonded, ball grid array and chip scale packages using a two-layer substrate, while Figure 6.25 shows a ball grid array with a conventional—not using build-up or other exotic materials—four-layer substrate.

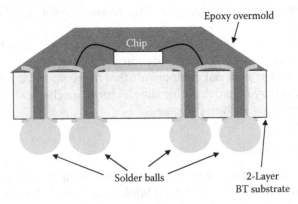

Figure 6.24 Cross section for a wire-bonded ball grid array or chip scale package using a two-layer substrate (not to scale).

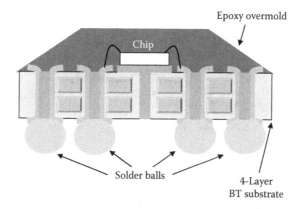

Figure 6.25 Cross section for a wire-bonded ball grid array using a four-layer substrate (not to scale).

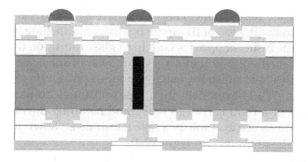

Figure 6.26 Cross section of a 2-2-2 buildup substrate (not to scale).

6.4.4.1   Microvias and high-density interconnect technology

To allow substrates to shrink further for chip-scale packages used in portable devices—like mobile handsets—microvia technology was employed. Microvia interconnects are formed using blind or buried vias, which often connect the outer layer with the layer directly underneath, thus saving real estate, because they do not extend through the entire board. An example is shown in Figure 6.26 of a 2-2-2 build-up substrate—two layers each of buildup on top and bottom with a conventional two-layer PCB board as the core.

Microvias fall under the broad category of high-density interconnect (HDI) technology. HDI substrates used lasers or photo-definition to create tiny vias—below 150-microns in diameter—rather than the traditional method of mechanically drilling them into the board. The advantages of high-density interconnect boards are several. There is the size/area/layer reduction while obtaining a higher wiring density. Another advantage

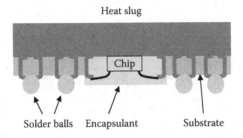

Figure 6.27 Cavity-down, thermally enhanced ball grid array (not to scale).

comes from electrical performance, with improved signal integrity and reduced electrical interference and cross-talk noise.

Currently, the outer layers of a multilayer substrate may consist of thin, built-up layers of metallization and dielectric, but the central two layers mimic that of a conventional printed circuit board. The buildup layers do not have glass reinforcement and often have a lower dielectric constant than that found with FR-4 or BT circuit boards. They include polyimides or photodielectric resins or epoxies with high glass transition temperatures.

Another method to obtain a very thin overall substrate thickness is by going with a tape substrate, usually made of polyimide tape with gold metallization. While extremely thin, tape substrates do require the use of a stiffener or a rigid carrier to withstand handling during the manufacturing process. An example of a tape substrate is shown in Figure 6.27 (which was already shown in Figure 2.4).

The options for buildup layer dielectric and via formation and subsequent metallization are numerous. Table 6.3 shows options available for dielectric materials, IVH (inner via hole) formation, and microvia metallization. Making various combinations—though they are not all interchangeable or can be matched with all of the possible options in the other columns—of the first three columns results in 21 different combinations, as listed in the fourth column, which were all available commercially at one time or another up through 2007.

6.4.5 Future developments

Substrate technology must evolve to match the performance of leading-edge chip designs and construction. As 45 nm and below technology nodes enter production, silicon chips fabricated with these ever-shrinking features may have higher frequency requirements than those presently available in current substrate technologies. Naturally, the pitch (50 μm and smaller) and via density requirements of these fine-featured devices must also be addressed and made compatible to the substrate. Issues such as electromigration and high-frequency performance must also be tackled.

Table 6.3 Types of Dielectric Materials, Inner Via Hole (IVH) Formation and Microvia Metallization Available to Create Buildup Layers, as of 2007

Dielectric Materials	IVH Formation	Microvia Metallization	Product Name
Photosensitive liquid	Photo via	Electroless copper	DV Multi-AD
Photosensitive dry film	Laser	Electroplated copper	IBSS-AAP/10
Polyimide flex circuit	Drill	Conductive paste	SLC
Thermal-cure dry film	Plasma	Bump connection	DV Multi-PID
Thermal-cure liquid	Screen print	Sheet copper bumps	HDI
Resin-coated copper (RCC)	Photo-lithography and Etch		MIKO-BU
FR-4 prepreg	Tool foil		CLLAVIS
Aramid prepreg			HITAVIA
Thermoplastic			Dycostrate
			SSP
			Fact-EV
			PPBU (HITAVIA)
			Laser-Via
			OrmeLink
			ALIVH
			MSF
			PALUP
			VIL
			BBIT
			NMBI
			Imprinted Circuits

Source: Adapted from Clyde Coombs, *Printed Circuit Handbook,* 6th Edition, McGraw-Hill Professional, New York, 23.2, 2007.

Other adaptations that may be required of substrate technology are compatibility with low-κ dielectric materials on the chip surface. Through-silicon vias (TSVs) may someday rival wire bonding or flip chip as a popular interconnect technology, and substrates and substrate design and bill of materials will likely need to change to meet this new technology's technical requirements.

Other desired improvements include lower moisture absorption and better warpage control for improved mechanical reliability. Though not required at this time, a substrates' bill of materials may need to adhere to "green" and lead-free environmental mandates in the future, which will

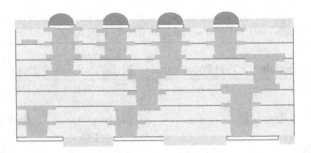

Figure 6.28 Cross section of a thin-core or no-core substrate (not to scale).

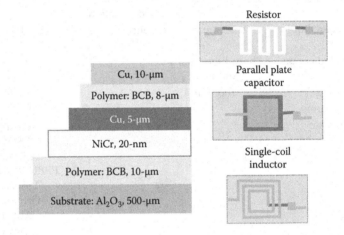

Figure 6.29 Representation of integrated passive components using thin-film technology.

surely require changes to the materials set. Changes to bonding materials, whether lead-free solder bumps or copper wedge bonds, may also require new bonding lead metallization finishes.

Another development is thin-core or coreless substrates. As already shown in Figure 6.26, the buildup layers still envelop a conventional circuit board sandwich. By going to a thin-core or no-core (Figure 6.28), the substrate thickness is greatly reduced, and consequently, so is the overall package profile. Here, "thin core" could mean going down to a thickness of 600 μm or 400 μm or below, compared to the current industry standard of 800 μm for the core of a high-performance substrate. Another benefit from minimizing or eliminating the core may come from shorter electrical path lengths providing incremental or better electrical performance.

Finally, development into integrating passive components into the substrate has been well underway. The difficulties come from the physical limitations of shrinking the size and footprint of passive components, as they cannot be scaled down to submicron levels. Perhaps going to thin-film construction may help the physical limitations. Some examples are shown in Figure 6.29.

Bibliography

H. Abe, "Molding Compounds for Cu Wire PKG," *K&S Copper Summit Conference*, July 15, 2008.

S. J. Adamson, "Review of CSP and Flip Chip Underfill Processes and When To Use the Right Dispensing Tools for Efficient Manufacturing," GlobalTRONICS Technology Conference, 2002.

M.R. Alam, A.B.M.A. Asad, M. Rahman, and K.S. Lee, "An Automated Mold Design System for Transfer Molding Process," *Proceedings of the International Conference on Mechanical Engineering 2007 (ICME2007)*, 29–31 December 2007.

K. Barrett, "More Things Come in Small Packages," *Electronic News*, September 13, 1999.

K. Barrett, "Flip-Chip on the March," *Electronic News*, January 15, 2001.

R.C. Benson, Dawnielle Farrar, and Joseph A. Miragliotta, "Polymer Adhesives and Encapsulants for Microelectronic Applications," *Johns Hopkins APL Technical Digest*, vol. 28, no. 1, 58–71, 2008.

J. Cannella, "Flip Chip Underfill Processing," *empfasis* (A publication of the National Electronics Manufacturing Center of Excellence), June 2006.

G. Carson and M. Todd. "Underfill Technology: From Current to Next-Generation Materials," *Advanced Packaging*, May 2006.

A.S. Chen, H.S. Chio, and R.H.Y. Lo, "Reliability Evaluation of Chip-On-Board (COB) Encapsulants and Substrate Interactions," in *SEMICON Taiwan 96—Packaging Seminar*, pp. 63–70, Taipei, Taiwan, September 24, 1996.

A.S. Chen, W.J. Schaefer, R.H.Y. Lo, and P. Weiler, "A Study of the Interactions of Molding Compound and Die Attach Adhesive, with Regards to Package Cracking," presented at *44th ECTC*, Washington, DC, 115–120, May 1–4, 1994.

S.S. Chen, J.E. Schoenberg, and S. Park, U.S. Patent 4, 975,221: High Purity Epoxy Formulations for Use as Die Attach Adhesive, December 4, 1990.

B. Chylak and I.W. Qin, "Packaging for Multi-Stack Die Applications," *Semiconductor International*, June 1, 2004.

B. Chylak, S. Tang, L. Smith, and F. Keller, "Overcoming the Key Barriers in 35 mm Pitch Wire Bond Packaging: Probe, Mold, and Substrate Solutions and Trade-offs," *SEMI Technology Symposium: International Electronics Manufacturing Technology (IEMT) Symposium*, July 17–18, 2002.

C. Coombs, *Printed Circuit Handbook,* 6th Edition, McGraw-Hill Professional, New York, 23.2, 2007.

B. Dipert, "Silicon Contends with Stuffed and Shrinking Packages," *EDN*, 49–58, June 13, 2002.

W. Eslinger, "Electromigration-Induced Failures in Plastic Encapsulated IC Packages," *Semiconductor International*, June 1, 2008.

R.H. Estes, "A Practical Approach to Die Attach Adhesive Selection," *Hybrid Circuit Technology*, June 1991.

R.H. Estes and F.W. Kulesza, "Current Technology of Die Attach Materials—Epoxy or Polyimide," *International Journal for Hybrid Microelectronics*, vol. 5, no. 2, November 1982.

C.M. Garner, F. Hua, N. Vodrahalli, A. Dani, T. Debonis, R. Aspandiar, and G. Brist, "Package Technology Trends and Lead Free Challenges," *2005 TMS Annual Meeting—Lead-Free Technology Workshop*, San Francisco, CA, February 13, 2005.

G.B. Goodrich and J.G. Belani, U.S. Patent No. 4,518,735: High Temperature Stable Adhesive for Semiconductor Device Packages, Low-Cost Semiconductor Device Package and Process, May 21, 1985.

N. Hackett and D. Loskot, "Encapsulation and Modern Molding," *Advanced Packaging*, October 2005.

C.A. Harper, *Electronic Packaging and Interconnection Handbook*, McGraw-Hill Professional, New York, Chapter 6, 1991.

H. Holden, "HDI Technology's Influence on Signal Integrity," *EE Times*, September 28, 2001.

H. Holden, "HDI Via Structures Effect on PCB Design Flexibility, Constraints and Cost," *Printed Circuit Design and Fabrication*, November 1, 2007.

B. Howe, "What is RoHS?," *IDES*, December 2005.

Instron, *Flexure Test*: www.instron.us/wa/applications/test_types/flexure/default.aspx?ref=http://www.google.com/search

Instron, www.instron.com

Intel Corporation, *Packaging Databook*, Chapter 3, 2000.

International Technology Roadmap for Semiconductors, 2007 Edition, Assembly and Packaging chapter.

IPC J-STD-030, draft 7, December 2000.

J. Israelson, "Packaging Solves 'the Last Centimeter,'" *EDN*, 73–82, May 24, 2001.

S. Iwasaki, M. Iji, and Y. Kiuchi, European Patent Application No. 98121270.7: Epoxy resin composition and semiconductor device encapsulated therewith, September 11, 1998.

M. Iyer, "Emerging Trends in Advanced Packaging," *Semiconductor International*, June 1, 2009.

S.C. Johnson, "Flip-Chip Packaging Becomes Competitive," *Semiconductor International*, May 2009.

R.C. Lasky and F. Komitsky Jr., "Die Attach in Lead Frame Packages: Step 4," *Advanced Packaging*, April 2004.

F.L.A. Latip, A. Hassan, and R. Yahya, "Delamination and Void Analysis on Die Attach Epoxy of a QFN Package," *Solid State Science and Technology*, vol. 16, no. 2, 207–213, 2008.

J.H. Lau, *Ball Grid Array Technology*, McGraw-Hill Professional, New York, Chapter 3, 1995.

D.S. Lee, D.W. Lee, and H.W. Park, U.S. Patent No. 6,106,259: Transfer Molding Apparatus with a Cull-Block Having Protrusion, August 22, 2000.

W. Lee, *Internal Document for Siliconware Precision Industries Co., Ltd.*, August 19, 2009.

Y. Li, "Accurate Predictions of Flip Chip BGA Warpage," *Proceedings of the 53rd Electronic Components and Technology Conference*, 549–553, May 27–30, 2003.

F. Liu, Y.P. Wang, K. Chai, and T.D. Her, "Characterization of Molded Underfill Material for Flip Chip Ball Grid Array Packages," *Proceedings of the 51st Electronic Components and Technology Conference*, 288–292, 2001.

R.H.Y. Lo and A.S. Chen, "Unconventional Molding Compounds for Conventional Packages," in *Proceedings FOCUS '94 Expo and Conference*, San Jose, CA, August 30–September 1, 1994.

R. Lo and H.P. Takiar, U.S. Patent No. 5,617,297: Encapsulation filler technology for molding active electronics components such as IC cards or PCMCIA cards, April 1, 1997.

R.H.Y. Lo and C.-C. Wu, U.S. Patent 6,507,120: Flip Chip Type Quad Flat Non-Leaded Package, January 14, 2003.

D. Maslyk, M. Privett, and B. Tolen, "Using Underfill to Enhance Lead-Free Drop Test Reliability," *SMT*, May 2006.

B. Mekdhanasarn, A.S. Chen, and R.H.Y. Lo, "Evaluation of Low Temperature Snap-Cure Die Attach Materials," *IEEE Transactions CPMT Part B: Advanced Packaging*, vol. 17, no. 1, 91–96, February 1994.

P.C.F. Moller, J. Mewis, and D. Bonn, "Yield Stress and Thixotropy: On the Difficulty of Measuring Yield Stresses in Practice," *Soft Matter*, vol. 2, 274–283, 2006.

National Semiconductor, *Application Note 1126: BGA (Ball Grid Array)*, August 2003.

National Semiconductor Corporation, *Data Sheet: Semiconductor Packaging Assembly Technology*, August 1999.

L.T. Nguyen, R.H.Y. Lo, A.S. Chen, and J.G. Belani, "Molding Compound Trends in a Denser Packaging World II: Qualification Tests and Reliability Concerns," *IEEE Transactions on Reliability*, vol. 42, no. 4, 518–535, December 1993.

L.T. Nguyen, A.S. Chen, and R.Y. Lo, "Interfacial Integrity in Electronic Packaging," *1995 ASME International Mechanical Engineer Congress and Exposition*, EEP-vol. 11, 35–44, San Francisco, CA, November 12–17, 1995.

M.N. Nguyen and K.-C. Le, U.S. Patent No. 5,708,129: Die Attach Adhesive with Reduced Resin Bleed, January 13, 1998.

Y. Nishi and R. Doering, *Handbook of Semiconductor Manufacturing Technology*, CRC Press, Boca Raton, FL, Chapter 32, 2000.

N.M. Patel, V. Wakharkar, S. Agrhram, N. Deshpande, M. Pang, R. Tanikella, R. Manepalli, P. Stover, J. Jackson, R. Mahajan, and P. Tiwari, "Flip-Chip Packaging Technology for Enabling 45 nm Products," *Intel Technology Journal*, vol. 12, no. 2, 145–156, June 17, 2008.

Prismark Partners, *Presentation for Siliconware Precision Industries Co., Ltd.*, August 20, 2009.

P. Procter, "Mold Compound—High Performance Requirements," *Advanced Packaging*, October 2003.

G.A. Riley, "Introduction to Flip Chip: What, Why, How," *FlipChips.com*, October 2000.

M.F. Rosle, I. Abdullah, S. Abdullah, M.A.A. Hamid, A.R. Daud, and A. Jalar, "Effect of Manufacturing Stresses to Die Attach Film Performance in Quad Flatpack No-Lead Stacked Die Packages," *American Journal of Engineering and Applied Sciences*, vol. 2, no. 1, 17–24, 2009.

L. Roth and G. Sandgren, "Wire Encapsulation Improves Fine-Pitch Device Yield," *Semiconductor International*, October 1, 2004.

W.L. Schultz and S. Gottesfeld, "Frequently Asked Questions About PEM Reliability," *1998 PEM Consortium*, Orlando, FL, February 6–7, 1998.

M. Schwartz, ed., *Smart Materials*, chapter 15, CRC Press, Boca Raton, FL, 2008.

Siliconware Precision Industries Co., www.spil.com.tw/

R.D. Skinner, ed., *Basic Integrated Circuit Technology Manual*, ICE (*Integrated Circuit Engineering*), 1993.

S. Bakelite, *Presentation: R&D Activities—EMC Key Technologies*, January 2009.

S. Bakelite, *Presentation: Technologies for Cu Wire Bonded PKG*, March 2009.

Y. Sun, *PhD Dissertation—Study on the Nanocomposite Underfill for Flip-Chip Application*, Georgia Institute of Technology, December 2006.

S.J. Taylor and R.J. Ulrich, "Implementation of On-Bonder-Curing to Maximize Array Package Manufacturing Productivity," presented at *SEMICON West 2000—Third Annual Semiconductor Packaging Symposium*, San Jose, CA, July 11, 2000.

"The Story Behind the Red Phosphorus Mold Compound Device Failures," *Reliability Information Analysis Center (RIAC) Desk Reference*, Third Quarter 2006.

T. Thompson, "Dispensing, Encapsulating and Underfilling: Picking the 'Right' Equipment for the Job," *Chip Scale Review*, January–February 2006.

M. Todd, "Material Systems Enable High Density Packaging," *EMI Asia*, April 1, 2008.

M. Todd, "Flow-Over-Wire Materials Enable Die Stacking," *Semiconductor International*, December 1, 2008.

B. Toleno, "STEP 5: Advanced Underfill Technology," *Surface Mount Technology*, May 2008.

M. Topper and P. Garrou, "The Wafer-Level Packaging Evolution," *Semiconductor International*, October 1, 2004.

P. Totta, ed., *Area Array Interconnection Handbook*, pp. 452–499, Springer, Berlin, 2001.

K. Tsuda, "Molding Techniques Support Thin Gold Wires, Low-k Materials," *Semiconductor International*, July 21, 2008.

R.R. Tummala and S. Chapman, *Fundamentals of Microsystems Packaging*, 372–379, McGraw-Hill Professional, New York, 2001.

C.P. Wong and M.M. Wong, "Recent Advances in Plastic Packaging of Flip-Chip and Multichip Modules (MCM) of Microelectronics," *IEEE Transactions on Components and Packaging Technology*, vol. 22, no. 1, 21–25, March 1999.

T. Winster, C. Borkowski, and A. Hobby, "Wafer Backside Coating of Die Attach Adhesives," *Semiconductor International*, October 1, 2006.

J.J. Zhang and D.F. Baldwin, "The Latest in Underfill for Advanced Chip Assembly," *Circuit Assembly*, September 2003.

Z. Zhang, J. Lu, and C.P. Wong, "Double-Layer No-Flow Underfill Process for Flip-Chip Applications," *IEEE Transactions on Components and Packaging Technologies*, vol. 26, no. 1, 239–244, March 2003.

chapter seven

Metals

7.1   Lead frames, heat spreaders, and heat sinks

7.1.1   Objectives

- Describe the purpose and usage of lead frames.
- Explain the difference between heat spreaders and heat sinks.
- Convey the purpose and importance of lead frames in semiconductor packaging.
- Discuss the special role of heat spreaders and heat sinks.

7.1.2   Introduction

This section looks at the use of metallic lead frames as a carrier for semiconductor chips. Also presented is a brief discussion on the use of heat spreaders and heat sinks to increase a chip's and package's thermal dissipation when necessary for proper operation and long operating life.

7.1.3   Lead frames

Metal lead frame carriers have been used in semiconductor packaging almost since the beginning of the industry. Although hermetic packages often used ceramic substrates as their chip support, nearly all of the early plastic packages used lead frames: dual in-line, small outline, plastic leaded chip carrier, and plastic quad flat packs. Figure 7.1 shows an example of a thin quad flat pack lead frame. (The polyimide tape seen on the leads lends support during assembly processes.)

Lead frames serve several purposes and functions. As implied by the name, the *leads* in a lead frame provide physical framework and electrical connections from the integrated circuit to the printed circuit board and system and on to the outside world. The paddle in the center of each package site in a lead frame is where the chip is attached and rests, and may even serve as a grounding point if such an electrical purpose is needed for a particular device.

Figure 7.1 Segment of a lead frame for thin quad flat packs. (From Wikimedia Commons, attributed to user NobbiP.)

Though substrate-based packages have overtaken lead frame–based packages in technological advances for many applications since their introduction in the early 1990s, lead frame packages remain the work-horses of the industry and are widely used. As shown in Table 7.1 (also presented as Table 2.1), nearly 70% of the world's semiconductor packaging output in 2007 utilized lead frames of one sort or another.

7.1.4   Metals commonly used in lead frames and other components

Each of the metals used as a lead frame material has advantages and dis-advantages. The following sections describe their usage and benefits. The discussion here does not cover the various plating finishes applied to the bonding surfaces of the lead frame, but concentrates only on the base metal.

7.1.4.1   Copper

Copper has the highest electrical conductivity of any metal except pure silver, and thus is used extensively in electronics for this purpose. But that is not the only advantage that copper has to offer. The metal is also

Table 7.1 2007 Worldwide Integrated Circuit (IC) Packaging Units by Package Family

Package Type	Share, %	
SO	24.8	
TSOP	13.4	
SOT	7.6	
DIP	5.3	
DCA	7.7	
WLP	4.2	
FBGA/DSBGA	13.7	
BGA	4.4	
PGA	0.1	
QFN	5.4	
QFP	9.2	
CC	0.8	
DFN	3.5	
Total	100	151 billion units

Notes: SO, small outline package; TSOP, thin small outline package; SOT, small outline transistor; DIP, dual in-line (through-hole leads) package; DCA, direct chip attach; WLP, wafer-level packaging; FBGA, fine pitch ball grid array/DSBGA, die-sized ball grid array; BGA, ball grid array; PGA, pin grid array packages; QFN, quad flat no lead; QFP, quad flat pack; CC, chip carrier; DFN, dual flat no lead.

Source: Adapted from Sandra Winkler, "Trends in IC Packaging and Multicomponent Packaging," *IEEE SCV Components, Packaging and Manufacturing Technology Chapter,* January 22, 2009.

a stable platform to withstand the various package assembly processing steps, through both mechanical and thermal stresses.

Copper alloys are commonly used to form lead frames, being easily fabricated and amenable to cold-forming operations such as stamping and bending. Small amounts—in the range of a few weight percentage—of alloying elements such as iron increase strength for forming and handling without completely sacrificing the electrical and thermal conductivity of pure copper.

Copper and its alloys are also easily solderable, readily making electrical interconnections. Copper slugs are often used as heat spreaders or heat sinks, due to their aforementioned heat dissipation qualities.

7.1.4.2   Alloy42

Alloy42 is a commonly used alloy for semiconductor packaging lead frames. It consists of 42% by weight of nickel (Ni), and the remainder is iron (Fe). The alloy is considerably stronger than copper alloys popularly used in lead frames but at the cost of a magnitude lower thermal conductivity—around 170 W/m-K versus about 14 W/m-K. Also, Alloy42's coefficient of thermal expansion (about $5 \times 10^{-6}/°C$) is much more closely matched to that of the silicon chip (about $3 \times 10^{-6}/°C$) compared to copper lead frames (about $17 \times 10^{-6}/°C$) or the printed circuit boards (over $17 \times 10^{-6}/°C$).

Alloy42 was more commonly used among Japanese suppliers of semiconductors. The rest of the world preferred the heat dissipation characteristics of copper over higher strength.

7.1.4.3 Aluminum

Aluminum is known as a versatile metal combining tensile strength with less weight. The metal can be easily fabricated by a variety of processes into a number of shapes and forms. Aluminum is commonly used as a heat spreader and heat sink material.

7.1.5 Heat slugs, heat spreaders, and heat sinks

All three of these items serve similar purposes—to use a piece of metal to dissipate more heat away from the chip than the packaging materials can do so alone, in order to prevent the chip's thermal shutdown, or worse. In fact, most of the materials used in plastic packages are poor or mediocre thermal conductors, especially the molding compound.

Generally, aluminum or copper metals are used as heat spreader or heat sink materials. These two metals are good at moving heat isotropically, but due to high contact resistance, they tend to be inefficient in transferring heat away from components. More exotic materials like graphite can work as heat spreader material. Graphite has anisotropic heat transfer properties—excellent in the x-y plane but a poor conductor in the z-axis. Graphite is very lightweight but tends to be much more brittle than either aluminum or copper.

Still, the use of any of these items for heat dissipation in a plastic package was avoided except when necessary, due to the additional cost, weight, and manufacturing steps. In the past, heat spreaders and heat sinks were only commonly used in power packages, such as the transistor outline (TO)-220 form factor, seen in Figure 7.2. However, they have become more common as semiconductors, in general, grow larger and more complex with more circuitry and therefore generate more heat.

7.1.5.1 Heat slugs or spreaders

Heat slugs and spreaders contact the backside of a lead frame die paddle, or the backside of the silicon chip in the case of a flipped chip. In either case, the metal is exposed and not completely covered by molding

Figure 7.2 Pair of TO-220 packages. (From Wikimedia Commons.)

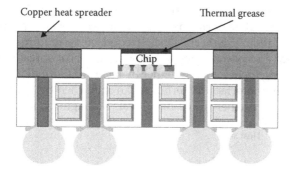

Figure 7.3 Example of a high-performance flip-chip ball grid array package (not to scale).

compound or other materials. They are integrated into the packages and are not removable. Their purpose is to radiate excess heat directly through a thermal path from the device. An example is illustrated in Figure 7.3—a high-performance flip-chip package. In the illustration, the gold-colored section above the chip is the heat spreader and the dark gray layer represents thermal grease that provides thermal contact between the chip and the heat spreader.

Another example is shown in Figure 7.4, which illustrates a leaded (small outline) package with a mechanically attached heat slug attached to the lead frame and exposed through the molding compound.

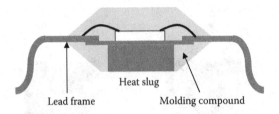

Figure 7.4 Example of a leaded plastic package with heat slug attached to die paddle (not to scale).

7.1.5.2   *Heat sinks*

Heat sinks are attached to the package after the fact, usually to an already embedded heat spreader. They are commonly large cubes of aluminum or copper formed in such a way to have multiple fins, in order to increase surface area as well as to facilitate the flow of air through and around the package in order to carry away the heat, usually by use of fans to facilitate forced-air cooling.

The use of heat sinks is most popularly associated with microprocessors, graphics processors, and computers, in general. As shown in Figure 7.5, the higher performance of microprocessors or graphics processors—and their gigahertz clock frequencies—requires solutions to dissipate heat and power, such as using both a heat spreader and a heat sink and having them joined together by layers of thermal grease or some type of thermal interface material. A fan may then be placed on top of the heat sink to further draw away heat. Managing a system's thermal

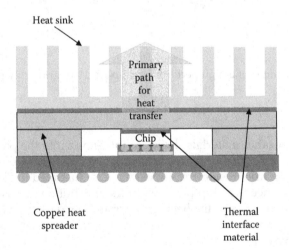

Figure 7.5 Example of a high-performance flip-chip ball grid array package with heat sink, intended for microprocessors (not to scale).

issues usually means using several cooling fans for the entire enclosure as necessary to prevent thermal overload on the key chip or chips. More extreme measures might involve the use of heat pipes—a sealed tube or container filled with a wicking material and a thermally conductive fluid—to pull heat out of the processor. Naturally, all these heat transfer solutions involve additional cost, weight, complexity, and manufacturing steps and tend to be avoided unless necessary.

7.1.6 Plating finishes

Typically, a lead frame would be spot plated with silver (Ag) on the die paddle area and on the lead tips, to provide oxidation protection and consistent surface condition for both the die attach step and for the wire bonding process. However, silver plating tended to have poorer adhesion to molding compound, which then led to delamination and other reliability issues.

In the past, on the external leads a lead-tin (Pb-Sn) solder would often be plated, to ensure good solderability conditions. With the advent of Restriction of Hazardous Substances (RoHS) and other environmental directives, eutectic lead-tin solder is no longer an acceptable surface finish, and alternatives have been sought. One surface finish in long use is nickel-palladium (Ni-Pd), sometimes with a gold (Au) flash finish for further stability. With nickel-palladium, the entire surface of the lead frame can be plated, and silver spot plating is not necessary. On the other hand, selection of molding compound may change, as formulations that adhere well to copper may not do so to nickel-palladium.

Still other plating finishes include pure tin, which is discussed further in Section 7.3.

7.2 Bonding wires

7.2.1 Objectives

- Describe the purpose of bonding wire and the metallic elements commonly used in the wire bonding process.
- Discuss the failure mechanisms associated with bonding wires.
- Briefly describe the testing techniques used to assess the quality of the wires and quality of their bonds.

7.2.2 Introduction

The use of very thin metal wires to provide electrical pathways from the integrated circuit to the package dates from the early days of the semi-

Table 7.2 Dopants Used in Gold Bonding Wires and Effect on Recrystallization Temperature

Gold Purity	99.999%	99.995%				
Dopant species	nil	Ag, Pd, Pt	Mg, Si, Ni	Co, Cu, Fe, Ga, Ge, In	Al, Be, Ca, Pb, Sn, Ti	
Recrystallization temperature[a]	150°C		150°C	150°C-200°C	200-300°C	Over 300°C

Source: Adapted from table 1 in Susumu Tomiyama and Yasuo Fukui, *Gold Bulletin*, 15(2), 1982.

[a] Degree of cold work is fixed at 99%. Temperatures quoted are for those measured for continuous annealing through a ring-type furnace.

conductor industry. The technique has been improved, refined, and automated over these many years but is essentially unchanged.

7.2.3   Bonding wires

Gold is now the standard for use in plastic semiconductor packages, and aluminum remains in use in ceramic and hermetic packages. Due to gold's rising cost, interest in copper as a less-expensive substitute is growing.

7.2.3.1   Gold

Gold has been the standard metal for bonding wires in plastic semiconductor packages for decades, due to its high environmental stability coupled with good electrical conductivity.

Gold wires come in 99.9999% purity (otherwise called "6N") or 99.99% (called "4N"). Dopants are added to improve various physical properties beneficial to thermosonic ball bonding. For instance, the use of certain dopants can raise the recrystallization temperature of gold wire, as shown in Table 7.2, which would be crucial to prevent uncontrolled changes to wire structure during the bonding process.

Dopants such as beryllium, calcium, and rare earth elements are known as interstitial dopants, which become tied up in the gold's grain boundaries and inhibit grain growth while increasing strength.

7.2.3.2   Copper

When gold prices reach very high levels, in the neighborhood of $1,000 per ounce—as it has again in recent years (Figure 7.6)—interest also reaches high levels to find a gold substitute. The most promising alternative to gold has been copper. Work on implementing copper wire bonding has waxed and waned through the years, as gold prices have risen and fallen. But this time it appears highly unlikely that gold prices will return

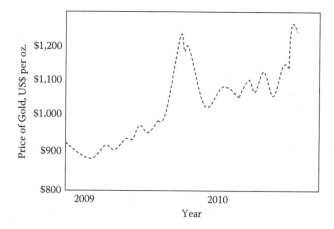

Figure 7.6 Price of gold during 2009 to 2010.

to recent low historical levels anytime soon, so copper wire bonding is finally gaining traction in the packaging industry.

Copper has several advantages: higher thermal and electrical conductivity (Table 7.3) with at least an order of magnitude lower material cost (Table 7.4). However, there are several disadvantages that have hampered copper's acceptance over the years: copper is much harder than gold and can damage the bond pads and underlying structures if care is not taken during processing, and copper requires the use of a forming gas—95% of N_2 plus 5% of H_2—during the bonding process to prevent copper oxidation.

Also during the bonding process, copper is not nearly as forgiving as gold. Bonding parameters must be set precisely in order to ensure good bond quality and subsequent reliability.

Copper wires generally used are 99.99% pure, doped with elements like calcium, silver, palladium, silicon, beryllium, phosphorous, manganese, and rare earths. The addition of dopants such as manganese and phosphorous improves ball bonding adhesion characteristics, as well as the FAB (or free air ball) shape. Elements like silver, palladium, platinum, and gold enhance the properties of manganese and phosphorous. Small amounts of silicon, beryllium, germanium, and others improve overall copper wire strength and elasticity. Finally, adding a little bit of calcium or some rare earth element, in addition to all of the previous dopants mentioned, can help control the length of the heat-affected zone and allow for low looping heights.

To date, the greatest usage of copper wire bonding has been with power semiconductors and discrete devices, which required very thick gold wire for current load-carrying capabilities. Switching to copper saves both considerable cost as well as improves electrical performance.

Table 7.3 Physical Properties of Copper and Gold Wire

	Copper Wire	Gold Wire
Electrical resistivity (micro-Ohm/cm)	1.7	2.3
Free air ball (FAB) hardness (Hv)	~90	~60
Ball bond hardness (Hv)	~128	~80
Tensile strength (gms)	8–15	10–15
Elongation (%)	8–16	2–6

Source: Adapted from table 1 in Sheila Rima C. Magno, Jean Ramos, Eduardo Pecolera, and Chris Stai, "Copper as a Viable Solution for IC Packaging," *Circuits Assembly*, February 1, 2008.

Table 7.4 Cost Comparison of Gold Wire versus Copper Wire

Wire Diameter (µm)	Cost of Wire per km			
	Gold at $800	Gold at $900	Gold at $1,000	Copper
20	186	205	225	40
25	274	304	335	40
30	381	425	469	40
38	593	664	734	40
50	1005	1127	1249	40
75	2224	2498	2772	40

Source: Adapted from Amy Low and Jack Belani, "Is Cu Wire-Bonding for Real?" presented at *IEEE-CPMT-SV Lunch Meeting*, April 23, 2009.

But again, with the continued high levels for the price of gold, the use of copper wire bonding is rapidly spreading to logic and higher pin count devices.

7.2.3.3 Aluminum

Aluminum wire is rarely used in plastic semiconductor packaging as it is prone to corrosion, and molding compound is obviously not impervious to moisture. However, it is commonly used in ceramic and hermetic packages, where it is employed in wedge-to-wedge bonding from chip to metal lead frame or hermetic substrate.

7.2.3.4 Other

Over the years, there have been attempts to use other metallic wires or alloys or even bimetallic wires—platinum-coater copper wire is the most recent example. So far, none have been particularly successful, whether due to technical or economic reasons.

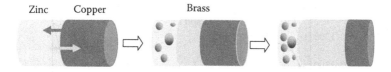

Zinc Copper Brass

Figure 7.7 Example of the Kirkendall effect using zinc and copper.

7.2.4 Kirkendall effect

The Kirkendall effect is named after Ernest Kirkendall who was an assistant professor at Wayne University from 1941 to 1946, during which time he wrote his now-famous paper titled "Zinc Diffusion in Alpha Brass," coauthored with Alice Smigelskas and published in 1947. Essentially, the Kirkendall effect describes that the diffusion of two solids into each other is not done at equal rates with the atoms of one replacing the atoms of the other in any one-to-one correlation. Instead, one solid will have a faster diffusion rate than the other, and the mixture/alloy of the two will grow into the faster diffusing material. Unfilled voids are left behind in the faster diffusing solid and eventually coalesce into large pores. Figure 7.7 shows an example of zinc and copper joined together. As diffusion occurs and a zinc-copper alloy is created, the zinc is consumed by the brass (the faster moving species), leaving behind larger and larger voids.

7.2.4.1 Gold-aluminum intermetallics and Kirkendall effect

It is well known that gold ball bonds on aluminum bond pads undergo the Kirkendall effect with the passage of time, especially when combined with elevated temperatures, producing gold-aluminum intermetallics. The high-temperature storage life reliability test (refer to Chapter 4) is aimed at eliciting failure modes based on this phenomenon.

There are five kinds of gold-aluminum intermetallics as shown in the phase diagram in Figure 7.8: Au_5Al_2 (tan-colored), Au_4Al (tan), Au_2Al (metallic gray), $AuAl$ (white-colored), and $AuAl_2$ (deep purple). The most well-known of these intermetallics is the deep purple–colored $AuAl_2$ that is named *purple plague*, as it was originally believed this particular intermetallic phase was responsible for bond failures. However, later studies showed that the Au_5Al_2 is the phase associated with bonding fails, as this intermetallic grows much faster than the other four phases.

Generally, as the intermetallic layers grow into the gold ball bond, they cause Kirkendall voids in the gold bulk. The primary failure mode for gold-aluminum bonds via the Kirkendall effect is a weakened interface between the gold-aluminum intermetallic area and the gold ball bond. After 1000 hours of high-temperature storage life at 150°C, there is often a great deal of Kirkendall voiding in the gold, as well as some level of physical separation between the gold and intermetallics (gold wire used was

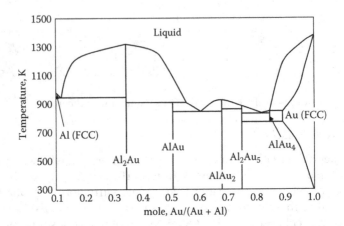

Figure 7.8 Aluminum-gold phase diagram (not to scale).

25 μm 4N type). Shear test failure mode is often through the intermetallic layer, though sometimes through the Au ball. There was no evidence of ball lift in the Jalar paper, in spite of severe intermetallic growth.

Although the Kirkendall voids are readily apparent in the gold, it turns out the aluminum on the bond pad is completely consumed as the process progresses, being such a thin layer to start with (approximately 1-μm thick).

7.2.4.2 *Kirkendall effect for copper wire bonding on aluminum bond pads*

It would be expected to see the Kirkendall effect with copper ball bonds on aluminum bond pads but this turns out not to be the case. The reason behind the difference is that radii of aluminum and gold atoms are nearly equal, but copper differs from the aluminum atom significantly, which results in a misfit of 10.5%. This hinders the aluminum atom movements in the copper ball bond and prevents solid solubility, which then inhibits the formation of copper aluminide.

7.2.5 *Heat-affected zone phenomenon in bonding wire*

The phenomenon of a heat-affected zone (HAZ) in metal joining is not unique to semiconductor assembly and manufacturing. It is instead a well-known phenomenon from welding metals in general. Figure 7.9 shows an example of the HAZ areas generated from welding.

The HAZ refers to the metallurgical changes seen in the volume of metal directly adjacent to the weld zone, where the metals saw elevated temperatures but not high enough temperatures to actually melt. However, the high temperature excursions in the HAZ region often alter

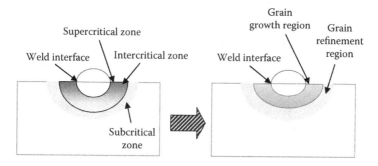

Figure 7.9 Heat-affected zone in welding.

the structure and even the material composition, often negatively. For instance, when welding steel, too much heat could lead to a HAZ area that has low impact strength (i.e., brittleness) due to recrystallization and coarse grain growth.

7.2.5.1   How is the heat-affected zone created?
Much the same effects are seen in gold bonding wire after the ball is created at the tip of the wire through electronic flame-off (EFO). The steps of thermosonic bonding by gold wires were shown in Figure 1.1 and are repeated in Figure 7.10. The heat generated by EFO travels a certain distance up the wire away from the ball and changes the grain structure and, by extension, the mechanical properties in that region.

7.2.5.2   Effect of heat-affected zone on loop height
As a general rule, having a long HAZ length precludes a low loop height, as it is unwise to bend and loop the wire within its HAZ. For instance, a high-strength gold wire has a shorter HAZ length than a standard 4N type, which should translate into a lower looping height.

7.2.6   Other reliability issues

The main reliability issues associated with gold wire are at the bond pad interface—Kirkendall effect, intermetallic growth, and phases—very pure gold is essentially a "noble" metal and normally does not react to its environment. On the other hand, metals like copper and aluminum tend to be much more reactive and therefore subject to environmental degradation.

7.2.6.1   Copper wire bonding and corrosion
Copper in general is more subject to oxidation and corrosion. Studies have found that impurities in packaging materials, especially with halogen

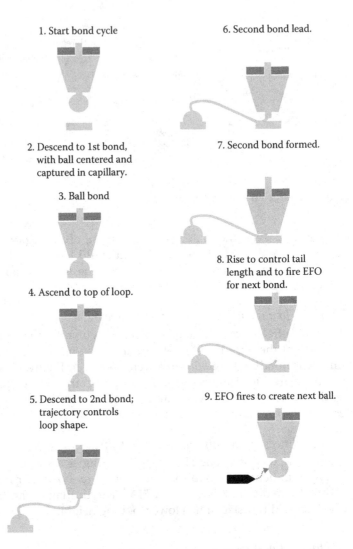

1. Start bond cycle

2. Descend to 1st bond, with ball centered and captured in capillary.

3. Ball bond

4. Ascend to top of loop.

5. Descend to 2nd bond; trajectory controls loop shape.

6. Second bond lead.

7. Second bond formed.

8. Rise to control tail length and to fire EFO for next bond.

9. EFO fires to create next ball.

Figure 7.10 Process steps in gold thermosonic wire bonding.

elements like chlorine, can degrade bond integrity at the bond pad in the presence of moisture. One solution is to use "green" molding compounds, which are required to have very low chloride levels.

7.2.7 Materials analysis

Below is a list of various methods to assess wire bond quality and post-failure analysis.

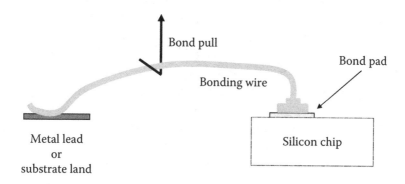

Figure 7.11 Test setup for bond pull.

7.2.7.1 *Visual inspection*

Visual inspection is used to monitor bond quality. Problems that could be observed are smashed bonds, small bonds, breaking wires, off-pad bonds, and bond pad cratering.

7.2.7.2 *Bond etching*

Bond etching is a destructive analysis technique that examines the condition of the bond pads and the structures underneath after bonding. With this technique, the ball (wire) and pad metallization are removed, and the underlying materials are visually examined for defects.

7.2.7.3 *Bond pull*

The bond-pull test is the primary method used for optimizing the bonding window and monitoring the bond quality. Figure 7.11 illustrates the setup. It should be understood that the pull test is influenced by package configuration and wire length. Test results include bond strength and failure mode.

7.2.7.4 *Ball shear tests*

Shear tests have long been used to determine the quality of a wire ball bonded to the bond pad. The amount of force required to push a ball bond off the pad was expected to correlate to the adhesion strength of the weld. An example of wire ball bond shear test setup is shown in Figure 7.12.

7.2.8 *Recent developments*

Below is an example of recent development work in the area of bonding wire and wire bonding technology.

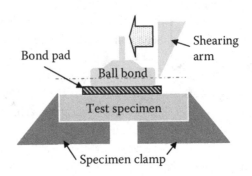

Figure 7.12 Cross section of ball bond shear test setup.

7.2.8.1 Copper wire bonding on nickel-palladium electroless plated bond pads

Recall the manufacturing issues encountered when performing copper wire bonding on standard aluminum bond pads. As noted before, there is the issue of "pad splash," when the aluminum gets pushed out from underneath the pad due to the excess bonding force and hardness of the copper. Thus, the process window with copper ball bonding on aluminum is rather narrow.

To address these and other issues, as well as increasing process tolerances, some semiconductor companies are electroless plating nickel on top of the bond pads. Nickel is 50% harder than copper and four times harder than aluminum. That additional hardness should reduce or eliminate any bond pad damage. Also, with the added hardness from nickel, there can be more leeway in the bonding parameters, allowing for higher ultrasonic energy usage.

A thin layer of palladium is plated on top of the nickel to prevent oxidation, which would degrade bonding performance. An addition of gold flash on the palladium may or may not be necessary for good bond strength and reliability.

7.3   Solders

7.3.1   Objectives

- Discuss the purpose and importance of solders in semiconductor packaging.
- Give a brief summary about the compositions of solders used in the electronics industry.
- Discuss the need for new solder alloys to replace eutectic tin-lead alloy in the face of RoHS and other environmental and "green" regulations.

7.3.2 Introduction

Solders and the process of soldering have been around a very long time, practically from the historic beginnings of the use of metals. Soldering provides the important function of joining two metallic surfaces mechanically, thermally, and electrically. Naturally, this purpose has turned out to be especially important in electronics components, where all three functions are necessary for the proper operation of a given electronics assembly.

7.3.3 Types of solders

Solders are generally alloys of two or three metallic elements. Elements commonly used in solder alloys are tin (Sn), lead (Pb), silver (Ag), bismuth (Bi), indium (In), antimony (Sb), and cadmium (Cd). The best-known solder alloys are tin-lead alloys, but there are many binary and ternary combinations using the listed metals.

Solder alloy selection for a given application depends on the following criteria:

- The solder alloy's melting temperature range with regards to the service temperature
- The alloy's strength in regard to service conditions
- Metallurgical compatibility to the surfaces being joined, including the possibility of intermetallic growth
- Environmental compatibility
- Wettability on surfaces to be joined

7.3.3.1 Lead-based

Soldering using lead-based alloys has been around for centuries, if not for millennia. The popularity of tin-lead binary solder alloys is likely due to the combination of two factors: relatively low processing temperature combined with acceptable mechanical properties. The two elements have complete liquid miscibility and partial solid miscibility, as shown in their phase diagram, given in Figure 7.13. The eutectic point is at 63% by weight of tin at a temperature of 183°C.

The drawbacks and health hazards of elemental lead are well known, and general concern for the environment has led to regulations leading to discontinuing its use, led by the European Union (EU) and Japanese governments. This, of course, has had—and still is having—a large impact on semiconductor packaging technology and has required major changes to decades of common practice, both in the bill of materials and procedures.

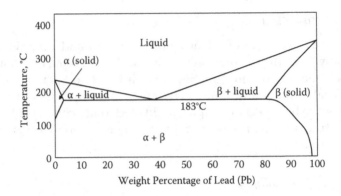

Figure 7.13 Lead-tin phase diagram (not to scale).

Now, high-lead (Pb)-content solders continue to be used in semiconductor packaging, for wafer bumping in a flip chip. This purpose is discussed further in Section 7.4.

7.3.3.2 Lead-free

As already discussed in Chapter 5, Section 5.1, the advent of RoHS, Japan's and other nations' environmental regulations means most use of lead-based solders must be phased out, sooner or later. For example, the original European Union's RoHS directive restricts lead levels to no more than 0.1% by weight in homogeneous materials as part of electrical and electronic equipment.

There are currently exemptions of lead-alloy solder usage for within a semiconductor package, such as high-lead-content solder bumps on a flipped die that is subsequently underfilled or encapsulated. Also, there are exemptions for high-performance applications, such as aerospace or medical equipment for life support or telecommunications equipment. However, at least some of those exemptions are expected to go away eventually, and lead-free alternatives will be required in the future, though not as of the year 2011.

After many years of research and development, several solder alloys have emerged to meet the combination of requirements—manufacturability, cost, availability, and reliability—as tin-lead replacements. Among the most popular in use are the tin-silver-copper tertiary alloys, commonly abbreviated as SAC alloys. Among the SAC alloys, one of the preferred alloys is 96.5% tin, 3.0% silver, and 0.5% copper, abbreviated as SAC305.

Another issue with lead-free solders that has far-reaching consequences is that these solders all require higher reflow temperatures than eutectic tin-lead alloy, at least 30°C to 40°C hotter at peak temperature—SAC305 has a melting range of 217°C to 220°C compared to eutectic tin-lead solder alloy melting point of 183°C. Therefore, the other packaging materials are subject to higher peak temperatures and elevated temperature profiles during assembly. As a consequence, heat resistance and package warpage have become more of a problem for the other parts that make up a given package, especially for molding compounds.

A third issue is that copper has proven to be more soluble in these lead-free solders compared to eutectic tin-lead. This could prove problematic in the case of wave soldering, where too thin a copper finish layer on the circuit board could end up nearly disappearing into the solder bath and affecting the final electrical performance of the system.

Another known risk associated with pure tin or high tin-content solders is tin whiskers that are crystalline threads of tin "growing" from surfaces where the tin (or high-tin-content solder) is the surface finish. Tin whiskers are electrically conductive and thus detrimental because they will cause short circuits. This phenomenon has been known since the 1940s and 1950s. Whiskering is also known to occur with other metals, such as zinc and antimony.

For unknown reasons, tin whiskers are much more unlikely to occur with alloy solders—like eutectic tin-lead. Thus, with the common usage of eutectic tin-lead solder in electronic assemblies for the past 50 years until the past 5 years, the phenomenon of tin whiskers was rarely ever encountered, since it had been "solved." By moving away from lead alloys, this issue has reemerged and new solutions are still being sought. Some studies have shown thick matte tin finishes may be less prone to whiskering than bright tin finishes or a thin matte layer. Another solution may be annealing the pure tin finish after plating for stress relief, as it is theorized that whiskering occurs to release stresses trapped in the grain boundaries. Further investigations into these and other fixes for tin whiskering continue.

Another phenomenon known to affect pure tin is called *tin pest*. Tin pest is much less publicized than tin whiskers, probably due to the fact it is only an issue at low temperatures, at below 13°C or 55.4°F, and happens rather rarely. When it does happen, the phenomenon manifests itself as a transformation from the commercially used, higher-density white (β) tin being slowly converted to the lower-density gray (α) tin over a long period of time (about 18 months). The transformation in density essentially causes the tin to fall apart and crumble.

Adding alloying metals to pure tin will retard or prevent this phenomenon from occurring. Of course, one of the elements that works very well at preventing tin pest turns out to be lead, but it also turns out silver and copper that are now commonly alloyed with tin for commercial lead-free solders, work well, too. The short list is shown in Table 7.5.

In any case, the literature indicates recreating tin pest at will is difficult at best. Nonetheless, it is a potential failure mechanism, even though the risk remains small.

Table 7.5 Alloying Metals to Prevent Tin Pest Phenomenon

Alloying Metal	Tin Pest Retardant Level	% Concentration for Effective Inhibition
Bismuth, Bi	Strong	0.3
Antimony, Sb	Strong	0.5
Lead, Pb	Strong	5.0
Copper, Cu	None to weak	? » 5.0
Silver, Ag	Weak to moderate	? > 5.0

Source: Adapted from Ronald C. Lasky, "Tin Pest: A Forgotten Issue in Lead-Free Soldering," given at *SMTAI*, September 2004, table 1.

7.3.3.3 Gold-based

Gold-based solder alloys are rarely, if ever, used in plastic semiconductor packages. They are used, typically as a die attach material, in hermetic and high-power/thermal dissipation applications.

For instance, gold-tin solder alloys are often employed as a die attach adhesive for microwave devices, laser diodes, and RF (radio frequency) power amplifiers. They are also seen in some forms of packaging for light-emitting diodes (LEDs). The eutectic alloy—80% Gold/20% Tin, which is shown in the phase diagram given in Figure 7.14—is well suited for these applications due to its high thermal conductivity (~57 W/m-K) and it does not require a flux in most cases to facilitate wetting. It is also lead-free, so it fulfills the environmental requirement. Coefficient of thermal expansion is reasonable and matches many other packaging materials, at $16 \times 10^{-6}/°C$. Finally, gold-tin has excellent resistance to corrosion. Drawbacks of gold-tin are its material

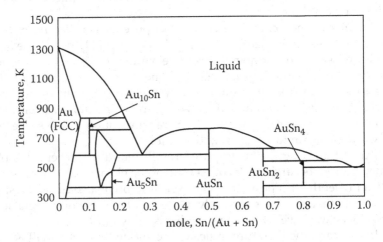

Figure 7.14 Gold-tin phase diagram (not to scale).

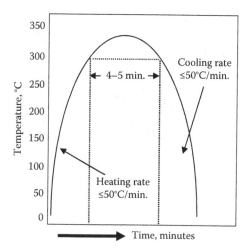

Figure 7.15 Gold-tin solder reflow profile (not to scale).

cost, especially as gold prices hover around $1,000 per ounce. Also, the device and product must be able to withstand an excursion to just over 300°C during die attach/reflow process, as shown in Figure 7.15. Last, the surfaces to be joined need a gold flash, though nickel-coated surfaces may be acceptable.

Another gold-based solder used in semiconductor packaging is gold-silicon alloy, which is typically used as a die attach material in hermetic packages.

7.4   Wafer bumping

7.4.1   Objectives

- Describe the various technologies used to create metallized bumps on wafers for flip-chip attachment or wafer-level packaging.
- Discuss the bump structures and the nature of under-bump metallurgies.

7.4.2   Introduction

Wafer bumping is necessary to create an electrical and physical connection between a chip and a substrate in lieu of wire bonding. This method is used in flip-chip packaging and in wafer-level chip scale packaging. This section discusses the various structures and compositions of the bumps placed on bond pads used to mechanically and electrically connect a flipped die to a package substrate or even directly to a printed circuit board (PCB). An example of a typical flip-chip interconnect structure is shown in Figure 7.16.

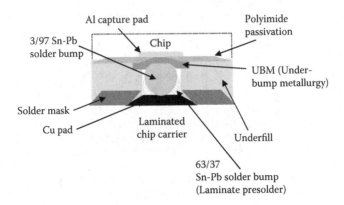

Figure 7.16 Example of a flip-chip interconnect.

The conductive bumps formed or placed on the bond pads fulfill several important electrical and mechanical functions. For one thing, it would be physically necessary to create a mechanical and electrical interconnect between the thin, recessed aluminum IC pads to a substrate land. Another issue is that connecting solder directly to aluminum is very tricky at best, but a thin metal layer would probably end up "leached" or dissolved by solder. A sandwich of various metallic layers, usually called the under-bump metallization or UBM, could prevent any material diffusion that would adversely affect the integrity of the connection. Another reason for the bumps is to insure a controlled gap between chip and substrate. If the chip and substrate were touching or nearly so, any CTE mismatch could result in high stress concentrations. By adding a gap, the bumps provide some stress relief.

7.4.3   Bump metallurgies

As implied by Figure 7.16, the most commonly used surface finish on a bump is a solder. The one illustrated in that figure is shown using a solder with high-Pb (lead) content in order to withstand the subsequent elevated temperature excursions during later assembly process steps. The evaporated high-Pb bump technology was developed by IBM ("C4") in the 1960s and was the first commercial use of flip-chip technology.

That is not to say other types of solders are not used. In common use—at least until the European Union's RoHS (Restriction of Hazardous Substances) directive went into effect in 2006—was eutectic tin-lead solder. Bump application techniques also varied, with electro- and electroless plating, and stencil printing, to name two.

In addition to solder, other processes have been developed, with varying degrees of success and commercial acceptance. These include gold

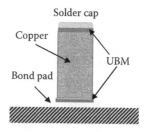

Figure 7.17 Tall copper pillar bump with solder cap.

stud bumping—derived from wire-bonding technology—and using anisotropic conductive films as a combination underfill and conductive path between bond pads and substrate lands.

Now with RoHS and other "green" initiatives, there is the move toward Pb-free solders and alternate bump metallurgies. For instance, Intel will pair up copper-plated bumps with tin-silver-copper solder for flip-chip interconnections at the 45-nm wafer technology node, though they are not the only ones to use this method: Analog Devices discussed using copper post bumps for its wafer-level packages in 2006. Figure 7.17 shows a version of the copper pillar bump with a high aspect ratio (i.e., a very tall column).

However, unlike eutectic tin-lead and high-lead solders, there is little long-term reliability history available, and so their long-term behavior out in the field is not really known.

7.4.3.1   "C4"

The Controlled Collapse Chip Connection—otherwise known as C4—process was developed by IBM in early 1960s. The process required evaporated under-bump metallurgy (or UBM) layers upon the bonding pads, made up of successive layers of chromium, chromium plus copper, copper, and topped with gold. Once the UBM layers were completed, high-lead solder was evaporated on top to form the bumps.

This process has a long manufacturing history and good reliability. The evaporation process allows for good control of alloy composition and uniform bump heights. The drawbacks are that the only solders that can be used are binary and high lead-based. The process is also not easily scaled up to large-sized wafers and requires a high capital investment with slow through-put.

An illustration of the C4 process is shown in Figure 7.18.

7.4.3.2   Electroplating

The following are detailed descriptions of the general steps involved in solder bumping wafers. An illustration of the process flow for electroplating bumps is shown in Figure 7.19.

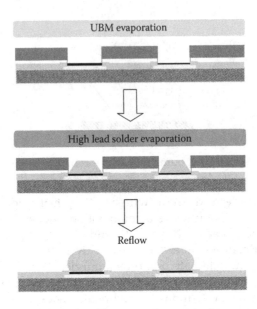

Figure 7.18 "C4" (Controlled Collapse Chip Connection) process steps.

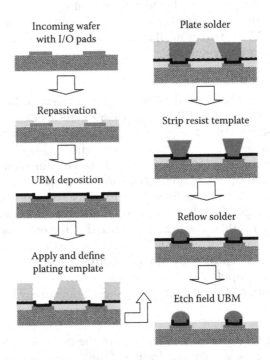

Figure 7.19 Process flow for electroplating solder bumps.

7.4.3.2.1 Wafer pretreatment. For the incoming wafer, plasma cleaning pretreatment is necessary to remove any oxides and organic residues. Plasma also roughens up the original passivation and bond pads to improve adhesion to the BCB (benzocyclobutene) as the dielectric material for the repassivation process. Repassivation is sometimes necessary to shrink the bond pad opening; otherwise, too much solder would be needed to create a bump with the proper standoff height for high reliability.

7.4.3.2.2 Benzocyclobutene passivation layer. The benzocyclobutene (BCB) undergoes coating, mask alignment and exposure, and development in order to open up the new bond pad openings.

7.4.3.2.3 Under-bump metal deposition. The under-bump metal is a key component in the solder bump structure with many key features, such as providing a barrier to the solder diffusion and having good adhesion to both the solder and to the initial aluminum bond pad. A common under-bump metal composition is Al/NiV/Cu (aluminum, nickel-vanadium, and copper).

7.4.3.2.4 Under-bump metal photo-development and removal process. Photo resist is next coated on the wafer, then mask alignment, exposure through the mask, and photo resist development to define the under-bump metal location and coverage. A wet etching process is applied to etch away any under-bump metal not covered by the remaining photo resist. The remaining photo resist is then removed with a solvent after under-bump metal etching.

7.4.3.2.5 Solder deposition. Next, the plating template is applied and defined with resist, and the solder is plated on the exposed pads. The resist is then stripped away and the solder reflowed. Once the solder bump has been formed, the wafer is now ready to go through the flip-chip assembly and packaging process.

Advantages of electroplating technology include good process control and the ability to create very fine-pitched bumps. Disadvantages include higher start-up costs and technique with many delicate process steps, especially with the plating bath to ensure uniform solder alloy composition and other physical characteristics. However, when done correctly, yields can be very high.

7.4.3.3 Electroless (UBM) plating and screen/ stencil printing solder

Actually, solder plating is very difficult, if not impossible, to do with electroless plating. What usually gets plated is the nickel under-bump metal, with a gold flash added on top for added environmental protection. The process flow is shown in Table 7.6. Typically, stencil printing solder on top is the next step, but nickel under-bump metal can be used in other flip-chip processes, as illustrated in Figure 7.20. Those other processes are discussed in Section 7.4.3.5.

7.4.3.3.1 *Screen/stencil printing solder.* The top layer of solder is applied by stencil printing solder paste onto the bumps. The initial capital investment is much lower than with electroplating. However, stencil printing limits the bump pitches to be coarser than those possible with electroplating—though pitches are not that coarse, because stencil-printed area array bumps can go as low as 120-μm pitch.

7.4.3.4   Lead-free bumping metallurgies

Similar to the activities going on with lead-free solders in general, the major difference is the aforementioned copper pillar bumps with a small amount of solder at the pillar top for the actual bonding.

Table 7.6 Process Steps in Electroless Plating of Nickel-Gold Bumps

Backside coating of wafer
Aluminum cleaning
Zincate pretreatment
Electroless plating of nickel (Ni)
Immersion flash plating of gold (Au)
Backside coating removal

Source: Adapted from Thorsten Teutsch, Thomas Oppert, Elke Zakel, Eckart Klusmann, Heinrich Meyer, Ralf Schulz, and Jorg Schulze, in *Proceedings of the 50th Electronic Components and Technology Conference*, Las Vegas, NV, May 21–24, 2000, figure 5.

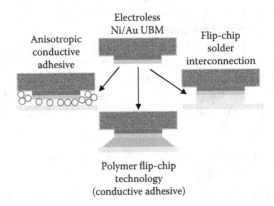

Figure 7.20 Many uses of electroless plated nickel-gold under-bump metallurgy.

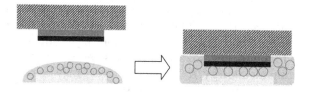

Figure 7.21 Flip-chip interconnection using anisotropic conductive adhesive.

7.4.3.5 *Alternative to solder bumping technologies*

There are a number of other techniques and metallurgies used to create bumps on bond pads. Some are used in production but usually for certain niches, and none are as widespread as solder bumps.

7.4.3.5.1 Gold stud bumping. A popular alternative is gold stud bumping. Gold stud bumping is essentially a gold ball bond without the wire. The wire is cut off close to the ball, leaving the stud on the bond pad. Advantages include using established equipment and technology—gold bonding wire and ball bonding equipment—and no need to redistribute bond pads on the chip or lay down an under-bump metal. Gold properties and behavior are also well understood, and the bumps offer high electrical and thermal conductivity. Disadvantages include substrate pad pitches that are often not as narrow as those found on the chips. Also, standoff height may be inadequate for reliability purposes.

7.4.3.5.2 Anisotropic conductive adhesive. Instead of using solder bumps or gold stud bumps, flip-chip electrical conductivity can be achieved with an organic polymer matrix embedded with tiny conductive pellets. The pellets clump together and provide electrical connection when the adhesive is pressed together between the chip and substrate. Some standoff height is given by electroless plating nickel (and gold flash) under bump metal on the bond pads. How the technology should work is shown in Figure 7.21.

7.4.4 *Under-bump metallurgy*

As already alluded to in the previous section, the purpose of under-bump metallurgy is to provide a metallurgy that will protect the chip and adheres well to the bond pad, and at the same time adheres well to the solder bump.

The under-bump metallurgy is usually made up of several layers of different metal alloys, with their functions described by the layers' names. The "adhesion layer" must adhere well to both the bond pad metal and the surrounding passivation, and provide a tough, low-stress mechanical and electrical connection. The "diffusion barrier" layer limits the diffusion of solder into the bond pad. The "solder wettable" layer offers an easily wettable surface to the molten solder during assembly, for good bonding to the under-bump metallurgy. A "protective layer" may be required to prevent oxidation of the under-bump metallurgy.

7.4.4.1 Vacuum deposition

Vacuum deposition was the initial method IBM came up with when it developed its C4 bumping process. It is an expensive process to lay down sequential metal layers and is rarely used anymore.

7.4.4.2 Electroplating

Electroplating is a well-established process used in various industries and for PCBs more specifically. IBM eventually replaced vacuum deposition with electroplating for its C4 solder bump process.

The steps required in electroplating can be complex, and cost is relatively high, but it offers good process control at fine dimensions, which makes electroplating the preferred choice for high input and output (I/O) fine-pitch bumping needs.

7.4.4.3 Electroless plating

This process involving the use of nickel as the plating material, commonly used in the automotive industry, has been applied to create flip-chip bumps. The process is easier to set up and maintain than electroplating, but there are some disadvantages. For one, the nickel bumps tend to be short, because there is no solder mask to control bump diameter in order to increase height. Also, the nickel still requires a solder surface finish for bonding, which is usually applied as a solder paste via stencil printing as mentioned previously. It turns out this lower-cost process works best on chips with low-to-medium bump density.

7.4.5 Technical issues

The issues concerning C4, tin-lead solders, and gold stud bumps are fairly well documented by this point. As already stated in the chapters for molding compound and other packaging materials, the potential problems surrounding lead-free and alternate "green" bump metallurgies will only come to light with time, unfortunately.

With solder bumps, they must be melted or reflowed to become fully stabilized. The main issue during reflow is oxide formation on the solder bumps, which would affect later processing and performance. Therefore, reflow usually occurs in a controlled atmosphere, usually of nitrogen or hydrogen, to prevent oxide from forming.

Another major issue is that of voids in the solder bumps, usually caused by poor wetting with the under-bump metal or excessive flux residue. Voids cause excessive collapse and therefore weak or nonexistent physical and electrical contact.

7.4.6 Future directions

With the RoHS and other environmental directives, the move to lead-free bumping will only continue. Recent developments and research point toward copper pillar bumps to provide the standoff height required for good reliability while minimizing the amount of solder necessary to make joint and connection.

Bibliography

C. Azzopardi, "Copper Wire Bonding on Nickel-Palladium-Gold Electro-less Plated Bumped Bond Pads," *K&S Copper Summit Conference*, July 15, 2008.

W.E. Bernier, "Flip Chip PBGA Assembly: Quality and Reliability Challenges," presented at *IMAPS*, October 2, 2008.

P. Biocca, "Lead-Free Wave Soldering—Some Insight on How to Develop a Process That Works," *EMSnow*, April 5, 2005.

E. Bradley, "Lead-Free Solder Assembly: Impact and Opportunity," presented at *53rd ECTC*, New Orleans, LA, 41–46, May 27–30, 2003.

D. Carrouge, H.K.D.H. Bhadeshia, and P. Woollin, "Microstructural Change in High Temperature Heat-Affected Zone of Low Carbon Weldable '13% Cr' Martensitic Stainless Steels," *Stainless Steel World*, 16–23, October 2002.

K. Chang, J.-Y. Lai, H. Pu, Y.-p. Wang, C.S. Hsiao, A. Chen, and R.H.Y. Lo, "Flip Chip Quad Flat No-Lead (FC-QFN)," *IWLPC 2005*, November 1, 2005.

F.-L. Chien, E. Ko, K. Chen, and A. Chen, "300 mm Wafer Bumping," *Future Fab International*, no. 13, July 8, 2002.

T. Collier and R. Young, "KGD Assembly Using Indium Converted Au Stud Bumps," *2006 KGD Packaging and Test Workshop*, September 10–13, 2006.

B. Chylak, J. Ling, H. Clauberg, and T. Thieme, "Next Generation Nickel-Based Bond Pads Enable Copper Wire Bonding," *Atotech.com*, 2009.

Computer History Museum, www.computerhistory.org/collections/accession/102677106

M. Conner, "Heat Spreaders and Fans Diffuse Hot Spots," *EDN*, pp. 55–58, March 17, 2005.

S. Davis, "Cooling Techniques Attack MPU Processing Heat," *Electronic Design*, October 23, 2002.

B. Dipert, "Silicon Contends with Stuffed and Shrinking Packages," *EDN*, pp. 49–58, June 13, 2002.

Freescale Semiconductor, *Application Note: Heatsink Small Outline Package (HSOP)*, AN2388, Rev. 1.0, December 2005.

K. Gilleo, "Tutorial 6—A Brief History of Flipped Chips," *FlipChips.com*, March 2001.

V. Gunaraj and N. Murugan, "Prediction of Heat-Affected Zone Characteristics in Submerged Arc Welding of Structural Steel Pipes," *Welding Journal*, 94-S–98-S, January 2002.

M. Hansen, *The Constitution of Binary Phase Diagrams*, 2nd Edition, McGraw-Hill, New York, 1958.

C.A. Harper, *Electronic Packaging and Interconnection Handbook*, McGraw-Hill Professional, New York, Chapters 1.3.5, 6, and 6.2, 1991.

J. Harris and E. Rubel, "Tough Thermal Apps Drive AuSn Die Attach," *Semiconductor International*, October 1, 2007.

V. Ho, S. Brown, R. Haythornthwaite, and D. Scansen, "Get Ready to Get the Lead Out," *Fabless Forum*, vol. 11, no. 3, September 2004.

A. Huffman, "50 Micron Pitch Flip Chip Bumping Technology: Processes and Applications," Presentation for *IEEE Components, Packaging and Manufacturing Technology Society, SCV Chapter*, September 13, 2006.

Indium Corporation, *Application Note: Gold Tin—The Unique Eutectic Solder Alloy*.

Intel Corporation, *Packaging Databook*, Chapter 3: Alumina and Leaded Molded Technology, 2000.

Intel Corporation, White Paper: RoHS Compliance at Intel, May 2006.

International Technology Roadmap for Semiconductors, 2007 Edition, Assembly and Packaging chapter.

ISSI, *Enhancing Long-Term Reliability with Copper Leadframes*, Application Note, 2008.

J. Jackson, P. Lomibao, and A. Jacobe, "Board Level Reliability of Wafer Level Chip Scale Packages with Copper Post Technology," Presentation for *IEEE Components, Packaging and Manufacturing Technology Society, SCV Chapter*, December 13, 2006.

A. Jalar, M.F. Rosle, and M.A.A. Hamid, "Effects of Thermal Aging on Intermetallic Compounds and Voids Formation in AuAl Wire Bonding," *Solid State and Technology*, vol. 16, no. 2, 240–246, 2008.

E. Kirkendall and A. Smigelskas, "Zinc Diffusion in Alpha Brass," 1947.

R.C. Lasky, "Tin Pest: A Forgotten Issue in Lead-Free Soldering," presented at *SMTAI*, September 2004.

J. Lee, M. Mayer, Y. Zhou, S.J. Hong, and S.M. Lee, "Tail Breaking Force in Thermosonic Wire Bonding with Novel Bonding Wires," *Materials Science Forum*, vols. 580–582, 201–204, 2008.

W. Lee, *Summary—Copper Wirebond Status, revision 4*, Siliconware USA, Inc. Internal Report, August 19, 2009.

F.-J. Leu, R.-S. Lee, H.-C. Huang, R.H.Y. Lo, and C.-H. Day, U.S. Patent No. 6,166,435: Flip-Chip Ball Grid Array Package with a Heat Slug, December 26, 2000.

J. Ling, Z. Atzmon, D. Stephan, and M. Sarangapani, "Wire Bond Reliability—An Overview on the Mechanism of Formation/Growth of Intermetallics," *SEMICON Singapore*, May 5–7, 2008.

R.H.Y. Lo, Z.M. Yang, and A.S. Chen, "IC Plastic Package Thermal Enhancements from a Materials Perspective," *Proceedings of the 7th International SAMPE Electronics Conference*, pp. 553–563, June 20–23, 1994.

R.H.Y. Lo, B. Mekdhanasarn, and D.P. Tracy, U.S. Patent No. 5,691,567: Structure for Attaching a Lead Frame to a Heat Spreader/Heat Slug Structure, November 25, 1997.

R.H.Y. Lo, B. Mekdhanasarn, and D.P. Tracy, U.S. Patent No. 6,479,323: Method for Attaching a Lead Frame to a Heat Spreader/Heat Slug Structure, November 12, 2002.

R.H.Y. Lo and C.-C. Wu, U.S. Patent No. 6,541,310: Method of Fabricating a Thin and Fine Ball-Grid Array Package with Embedded Heat Spreader, April 1, 2003.

R.H.Y. Lo and E. Tjhia, "Backsputtering Etch Studies in Wafer Bumping Process," *Solid State Technology*, pp. 91–94, June 1990.

A. Low and J. Belani, "Is Cu Wire-Bonding for Real?" at *IEEE-CPMT-SV Lunch Meeting*, April 23, 2009.

S. Rima C. Magno, J. Ramos, Eduardo Pecolera, and Chris Stai, "Copper as a Viable Solution for IC Packaging," *Circuits Assembly*, February 1, 2008.

D. Manessis, R. Patzelt, A. Ostmann, R. Aschenbrenner, H. Reichl, J. Wiese, and C. Modes, "Technological Advancements in Lead-Free Wafer Bumping Using Stencil Printing Technology," *EMPC 2005*, 427–433, June 12–15, 2005.

B. Mekdhanasarn and R.H.Y. Lo, U.S. Patent No. 5,773,876: Lead Frame with Electrostatic Discharge Protection, June 30, 1998.

B. Mekdhanasarn and R.H.Y. Lo, U.S. Patent No. 5,891,760: Lead Frame with Electrostatic Discharge Protection, April 6, 1999.

B. Mekdhanasarn, R.H.Y. Lo, Steve M. Ichikawa, and Abdul Rahim Ahmed, U.S. Patent No. 5,796,570: Electrostatic Discharge Protection Package, August 18, 1998.

G. Milad, "Is 'Black Pad' Still an Issue for ENIG?" *Circuit World*, vol. 36, no. 1, 10–13, 2010.

J.T. Moon, J.S. Hwang, J.S. Cho, and S.H. Kim, "New Materials for Bonding Wire," *SEMICON Singapore*, May 5–7, 2008.

S. Murali, N. Srikanth, and Y.M. Wong, "Fundamentals of Thermo-Sonic Copper Wire Bonding in Microelectronics Packaging," *Journal of Materials Science*, vol. 42, 615–623, 2007.

H. Nakajima, "The Discovery and Acceptance of the Kirkendall Effect: The Result of a Short Research Career," *JOM*, vol. 49, no. 6, 15–19, 1997.

NASA, Tin Whiskers, http://nepp.nasa.gov/WHISKER/

L.T. Nguyen, A.S. Chen, and R.Y. Lo, "Interfacial Integrity in Electronic Packaging," *ASME 1995—Application of Fracture Mechanics in Electronic Packaging and Materials*, EEP-vol. 11/MD-vol. 64, 35–44, 1995.

T. Oppert, "How to Grow a Flip Chip Bumping Service Business," *IMAPS Global Business Council*, June 21, 2004.

N.M. Patel, V. Wakharkar, S. Agrahram, N. Deshpande, M. Pang, R. Tanikella, R. Manepalli, P. Stover, J. Jackson, R. Mahajan, and P. Tiwari, "Flip-Chip Packaging Technology for Enabling 45 nm Products," *Intel Technology Journal*, vol. 12, no. 2, 145–156, June 17, 2008.

P. Preuss, "Hollow Nanocrystals and How to Mass Produce Them," *Science Beat– Berkeley Lab*, May 28, 2004.

I.W. Qin, "Wire Bonding Tutorial," *Advanced Packaging*, July 2005.

G. Reed, "Wafer Bumping: Is the Industry Ready?" *Semiconductor International*, October 1, 2004.

G.A. Riley, "Solder Bump Flip Chip," *FlipChips.com*, November 2000.

G.A. Rinne, "Tin Pest in Tin-Rich Solders," *Advanced Packaging*, November 2006.

K. Schischke and E. Jung, "The Lead-Free Challenge: Materials for Assembly and Packaging," *Semiconductor International*, August 1, 2004.

G.E. Servais and S.D. Brandenburg, "Wire Bonding—A Closer Look," presented at ISTFA'91, Los Angeles, CA, November 11–15, 1991.

J. Seuntjens, Z.P. Lu, R. Emily, C.W. Tok, F. Wulff, S.S. Aung, and S. Kumar, "Development of New Ultra-High Stiffness Gold Bonding Wire," *SEMICON West Advanced Packaging Technology Symposium*, July 16–20, 2001.

A. Shah, M. Mayer, Y. Zhou, S.J. Hong, and J.T. Moon, "In Situ Ultrasonic Force Signals during Low-Temperature Thermosonic Copper Wire Bonding," *Microelectronic Engineering*, vol. 85, 1851–1857, 2008.

V. Solberg, "Designers Guide to Lead-Free SMT," *IPC APEX EXPO*, Las Vegas, NV, March 29–April 2, 2009.

B. Swiggert, "Copper Wire Bonding: Panacea or Pandora's Box?" *K&S Copper Summit Conference 2008*, July 15, 2008.

L.C. Tan, "Copper Wire Bond Outlook," *K&S Copper Summit Conference 2008*, July 15, 2008.

T. Teutsch, T. Oppert, E. Zakel, E. Klusmann, H. Meyer, R. Schulz, and J. Schulze, "Wafer Level CSP Using Low Cost Electroless Redistribution Layer," *Proceedings of the 50th Electronic Components and Technology Conference*, Las Vegas, NV, May 21–24, 2000.

S. Tomiyama and Y. Fukui, "Gold Bonding Wire for Semiconductor Applications," *Gold Bulletin*, vol. 15, no. 2, 1982.

T. Uno, K. Kimura, and T. Yamada, U.S. Patent Application No. 20080061440: Copper Alloy Bonding Wire for Semiconductor Devices, March 13, 2008.

J. Vanfleteren, "Adhesive Flip-Chip Technology," *IMAPS Benelux Spring Event*, May 14, 2004.

S. Winkler, "Trends in IC Packaging and Multicomponent Packaging," *IEEE SCV Components, Packaging and Manufacturing Technology Chapter*, January 22, 2009.

H. Xu, C. Liu, V.V. Silberschmidt, S.S. Pramana, T.J. White, and Z. Chen, "A Re-examination of the Mechanism of Thermosonic Copper Ball Bonding on Aluminium Metallization Pads," *Scripta Materialia*, vol. 61, 165–168, 2009.

chapter eight

Ceramics and glasses

8.1 Objectives

- Briefly discuss the uses of ceramic and glass materials in semiconductor packaging.
- Describe the types of ceramic and glass materials used in semiconductor packaging, along with their advantages and disadvantages.
- Describe and illustrate some commonly used hermetic semiconductor packages.

8.2 Introduction

Ceramics and glasses used in electronic packaging are typically electrical insulators. They are not generally used in plastic semiconductor packaging but are rather used in hermetic packaging given their physical properties.

As stated in Chapter 1, hermetic packages are needed for high-reliability, high-performance applications such as military and aerospace applications. By definition, a hermetic seal prevents gases and liquids from penetrating the package and adversely affecting the integrated circuit. Also, because the package is made of ceramic or similar materials, the package can withstand higher operating and environmental temperatures than an equivalent plastic package.

Glasses are also rarely used in plastic semiconductor packages. Again, they are more typically employed in hermetic packages, either as a die attach material (silver-filled glass) or else used to seal ceramic lids to the package body (lead alkali borosilicate glasses), among other purposes.

8.3 Types of ceramics used in semiconductor packaging

As already stated, most ceramics are electrical insulators. A popular ceramic material used in electronic and semiconductor packaging applications is alumina (aluminum oxide or Al_2O_3) but others used include aluminum nitride (AlN), beryllium oxide (also known as beryllia or BeO), silicon carbide (SiC), and boron nitride (BN). Table 8.1 shows a comparison of various material properties for several ceramic materials.

Table 8.1 Material Properties of Ceramic and Other Materials Used in Electronic Packaging

Material	Density, g/cm^3	Coefficient of Thermal Expansion (CTE), ppm/ °C (20°C–150°C)	Thermal Conductivity, W/m-K	Bend Strength, MPa	Young's Modulus, GPa
Silicon, Si	2.3	4.2	151	—	112
Copper, Cu	8.96	17.8	398	330	131
Aluminum, Al	2.7	23.6	238	137–200	68
Alumina, Al$_2$O$_3$	3.98	6.5	20–30	300	350
Aluminum nitride, AlN	3.3	4.5	170–200	300	310
Silicon carbide, SiC	3.2	2.7	200–270	450	415
Beryllia, BeO	3.9	7.6	250	250	345
Gallium arsenide, GaAs	5.23	6.5	54	—	—
AlSiC (60-vol% SiC)	3	6.5–9	170–200	—	—
Kovar (Ni-Fe)	8.1	5.2	11–17	—	131
Cu-W (10-20% Cu)	15.7–17	6.5–8.3	180–200	1172	367
Cu-Mo (15-20% Mo)	10	7–8	160–170	—	313

Source: Adapted from Mark Occhionero, Richard Adams, and Kevin Fennessy, in *Proceedings of the Fourth Annual Portable by Design Conference, Electronics Design*, pp. 398–403, March 24–27, 1997, table 1.

8.3.1 Alumina

In the early days of the semiconductor and electronic packaging industry, aluminum oxide was found to be useful due to high thermal conductivity compared to other ceramics available at that time period, up to 20 times more thermally conductive.

Some of the later issues found with alumina include its somewhat high dielectric constant ($\kappa \sim 10$) and higher thermal expansion coefficient compared to silicon ($7 \times 10^{-6}/°C$ versus $4 \times 10^{-6}/°C$). Its dielectric constant is incompatible with high-frequency applications, and thermal expansion mismatch can prove problematic with larger-sized silicon chips. Also, modern power applications find alumina's level of thermal conductivity inadequate.

To be useful as substrate material, metallic conductors need to be bonded onto the ceramic surface. Metal traces are laid upon alumina substrates either through high-temperature firing or low-temperature thick-film processing. The advantage of high-temperature firing, usually with molybdenum or tungsten, is its moderate cost, but metallization tends to have a high resistivity. Low-temperature thick-film processing may have cost issues, but the final metallization resistivity is low. These processes are also applicable to other ceramics used in electronics.

8.3.2 Beryllia

Beryllium oxide, more commonly known as beryllia, has excellent thermal conductivity, at least 10 times higher than alumina, among other desirable physical properties. Its fundamental drawback is toxicity—beryllium and its compounds are very hazardous to human health, and extreme care is necessary in handling and machining beryllia to prevent toxic exposure to manufacturing staff. Thus, berryllia's use is generally very limited, often to applications where thermal conductivity is paramount and cost is secondary.

Beryllia processing and assembly techniques tend to be similar to those of alumina, though with much greater care to prevent toxic exposure.

8.3.3 Aluminum nitride

Although its thermal conductivity is not as good as beryllia, aluminum nitride does not have the health hazards associated with the former and has proven a popular choice for applications where good thermal conductivity is needed. However, there are other material issues associated with this compound. For one, it is much more difficult to fully sinter aluminum nitride to full density compared to alumina. If the part is not at full density, that means tiny voids or microcracks are present, which would compromise the part's integrity. And, the dielectric constant is similar to that of alumina, so usage for high-frequency applications would also be limited.

Another issue is chemical stability. Aluminum nitride is not a naturally occurring compound. If moisture is present, aluminum nitride can decompose to alumina and ammonia, by the reaction shown in Equation (8.1):

$$4AlN + 8H_2O = 2Al_2O_3 + 4NH_4 + O_2 \tag{8.1}$$

Finally, another consideration is a metal's adhesion to aluminum nitride. Many of the glasses used as a binder in thick-film pastes can reduce the compound to alumina by the following reaction shown in Equation (8.2):

$$2AlN + 3MnO_2 = Al_2O_3 + 3MnO + N_2 \tag{8.2}$$

The nitrogen gas released in the above reaction can cause blistering in the metallic film or a weak porous adhesion layer.

Therefore, processing and handling techniques tend to be very different from that of alumina or beryllia and possess their own difficulties.

8.3.4 Silicon carbide

Silicon carbide also has very high thermal conductivity and none of the toxicity issues of beryllia. A comparison of the thermal conductivities over temperature of the aforementioned ceramics is shown in Figure 8.1.

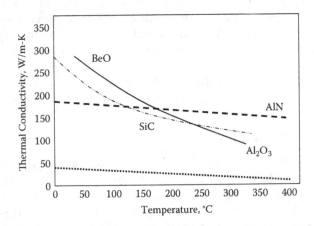

Figure 8.1 Thermal conductivity over temperature of ceramic materials used in electronic packaging.

However, silicon carbide has a relatively high dielectric constant ($\kappa = 40$) and is nominally an electrical conductor only within individual grains, while the intergranular phase acts as an insulator.

8.3.5 Boron nitride

The interest in boron nitride is due to its relatively elevated thermal conductivity, about twice that of alumina at 60 W/m-K. However, boron nitride cannot be metalized, nor sealed to, with a bonding material or technique; therefore, its use in semiconductor and electronic packaging is extremely limited.

8.4 Types of glasses used in semiconductor packaging

Glasses serve many purposes in semiconductor and electronic applications. They may act as insulators and passivation materials, or as bonding layers or package sealants, along with many other uses.

Glasses are noncrystalline solids, with the random structure of liquids frozen into place as the molten glass cooled. In other words, glass is a liquid at a temperature where it is so stiff that for all intents and purposes it is rigid. Table 8.2 gives some physical properties of glasses commonly seen in electronic applications. The viscosity data shown in Table 8.3 show the different temperatures associated with a given glass that show it becomes less rigid by some measure. For example,

Table 8.2 Material Properties of Glasses Used in Electronic Packaging

Glass	Thermal Expansion (0 to 300°C), ppm/°C	Density, g/cm³	Young's Modulus, GPa	Poisson's Ratio	Dielectric Constant (κ)
Soda lime	92	2.47	10.0	0.24	72
Borosilicate[a]	33–46	2.23–2.28	~65	~0.21	~4.7
High lead (Pb)	84	5.42	67	0.28	15.0
96% silica (SiO_2)	8	2.2	68	0.19	3.8
Fused silica (SiO_2)	5.5	2.64	71.7	0.16	3.8

Source: Adapted from Charles A. Harper, *Electronic Packaging and Interconnection Handbook,* McGraw-Hill Professional, New York, chapter 1.4, 1991, table 1.18.

[a] Varies by composition.

Table 8.3 Viscosity Properties of Glasses Used in Electronic Packaging

Glass	Strain Point, °C	Annealing Point, °C	Softening Point, °C	Working Point, °C
Soda lime	470	510	695	1005
Borosilicate[a]	435–515	480–565	710–820	1115–1245
High lead (Pb)	340	365	440	560
96% silica (SiO_2)	820	910	1500	—
Fused silica (SiO_2)	990	1050	1580	—

Source: Adapted from Charles A. Harper, *Electronic Packaging and Interconnection Handbook*, McGraw-Hill Professional, New York, chapter 1.4, 1991, table 1.18.

[a] Varies by composition.

- *Strain point*—the temperature at which strain is relieved in a few hours
- *Annealing point*—the temperature at which strain is relieved in a few minutes due to viscous flow
- *Softening point*—the temperature at which a given glass fiber will deform under its own weight
- *Working point*—the temperature at which an amount of glass can be easily shaped

"Liquids" glasses tend to have a wide range of compositions because they tend to act more like chemical mixtures. Different oxides and elements act as dopants in glasses to influence its physical properties, like coloration or softening agents.

Nonetheless, glasses are generally based on one of four oxides: SiO_2 or silica, B_2O_3, P_2O_5, and, more rarely, GeO_2. The first three oxides can be intermixed as glasses, along with the aforementioned dopants. Table 8.4 shows some possible compositions of glasses used in electronic packaging.

Table 8.4 Possible Compositions of Glasses Used in Electronic Packaging, by Weight Percentages

Glass	SiO_2	Na_2O	K_2O	PbO	B_2O_3	Al_2O_3
Solder seal	3		75	11	11	
Sealing	35		7	58		
Labware	80	3.5	0.5		14	2
96% silica	96	0.2	0.2		3	0.6
Fused silica	99.5					

Source: Adapted from Charles A. Harper, *Electronic Packaging and Interconnection Handbook*, McGraw-Hill Professional, New York, chapter 1.4, 1991, table 1.19.

Here in this section, the discussion will be limited to two of the types associated with packaging and assembly: silver-filled glass and lead alkali borosilicate glass.

8.4.1   Silver-filled glass

Silver-filled glass is a suspension of silver and low-softening temperature glass particles in an organic vehicle, becoming a paste. For die attachment, the paste is applied to the ceramic via a paste dispense system. After the paste is applied, the die is positioned within the dispensed pattern. Because of the high organic content of the paste, the silver-filled glass is carefully dried in a continuous furnace to remove the solvent. (The presence of solvent will lead to poor adhesion to the die.) This leaves behind a resin that binds the silver and glass particles until subsequent processing can soften the glass. After drying, the material is carefully heated to remove the binder; the heating rate being determined by the size of the chip.

Silver-filled glass, like gold-silicon eutectic solder alloy, is limited to those devices that can withstand elevated processing temperature and prolonged processing time. The silicon fabrication processes should be designed to withstand the die attach process. Because of the organic content—solvent and resin binder—of the silver-filled glass paste, its removal is necessary for good adhesion. And as stated already, the larger the chip, the longer is the drying/processing time for organic content removal.

8.4.2   Lead alkali borosilicate glass

These glasses—based on silica mixed lead, boron, bismuth, or zinc oxides—are widely used in sealing and adhesive applications due to their low melting and softening temperatures. They find extensive use in sealing ceramic or metal lids to packages to create a hermetic seal. In powder form, they are often added to thick-film pastes due to their adhesive properties.

Bibliography

E. Bogatin, "Ceramics Technology Top 10," *Semiconductor International*, January 1, 2003.

C.A. Harper, *Electronic Packaging and Interconnection Handbook*, McGraw-Hill Professional, New York, Chapters 1.4, and 6, 1991.

Intel Corporation, *Packaging Databook*, Chapter 3, 2000.

National Semiconductor Corporation, *Data Sheet: Semiconductor Packaging Assembly Technology*, August 1999.

M. Occhionero, R. Adams, and K. Fennessy, "A New Substrate for Electronic Packaging: Aluminum-Silicon Carbide (AlSiC) Composites," *Proceedings of the Fourth Annual Portable by Design Conference, Electronics Design*, 398–403, March 24–27, 1997.

E. Savrun, "Packaging Considerations for Very High Temperature Microsystems," *Proceedings of IEEE Sensors*, vol. 2, 1130–1143, June 12–14, 2002.

J.W. Soucy and T.F. Marinis, "Aluminum Nitride Chip Carrier for Microelectronic Sensor Applications," *Proceedings of the MRS Fall Meeting*, vol. 741, 2002.

section four

The future

chapter nine

Trends and challenges

9.1 Objectives

- Look at upcoming wafer fabrication developments and how they will affect semiconductor packaging technology.

9.2 Introduction

This chapter looks at the continued scaling of silicon complementary metal oxide semiconductor (CMOS) chips and the issues imposed on assembly and packaging requirements by the new materials set used to create these integrated circuits. There is also some discussion of what may be required of semiconductor packaging once Moore's Law can no longer be sustained.

9.3 Copper interconnects and low-κ dielectric materials

According to the 2007 update to the Interconnect chapter, ITRS (International Technology Roadmap for Semiconductors) noted that as far back as 1994 it was projected that new interconnect and dielectric materials would be necessary to replace aluminum and silicon dioxide, and the first implementation of copper-containing chips was imminent in the 1997 edition. However, widespread adoption of copper for the interconnect combined with low-κ dielectric materials did not happen until the mid-2000s, mostly due to issues with the new insulator materials—copper interconnect adoption actually preceded that of new dielectric materials.

9.3.1 Copper interconnects

Metallic interconnects are conductive traces on an integrated circuit, whose purpose is to distribute clock and other signals and to connect the power/ground to the various circuit and system functions on a chip.

The standard metallization for interconnects used for silicon CMOS was based on aluminum for years, if not decades. But to keep up electrical performance requirements as feature sizes shrank, it was determined that

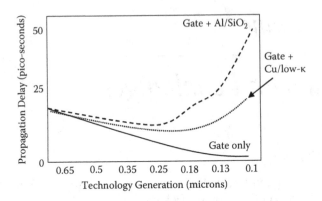

Figure 9.1 Dependence of propagation delay on a given technology generation.

there would be a need to switch to copper once the wafer process technology node dropped below 0.35 μm, as effects from inherent interconnect properties overtook other factors, such as due to design. An example comparing the electrical performance between aluminum plus silicon dioxide versus copper plus low-κ (κ being dielectric constant) insulator materials is shown in Figure 9.1. Until the 0.18-μm process technology node was reached, though, other materials and process changes were employed to address various issues at each scaling node, as shown in Table 9.1.

Table 9.1 Interconnect Innovations and Their Drivers

Technology	Node, μm	Technology Driver
AlSi alloy	1.0	Contact reliability (leakage/spiking)
AlSiCu alloy	0.8	Line reliability (electromigration)
TiN/TiW barrier layer	0.5	Contact reliability (R_c, spiking)
Tungsten (W)-plug	0.5	Scaling—straight sidewalls in contacts and vias (step coverage)
TiN-AlCu-TiN metal lines	0.5	Reliability—hillocks, top ARC provision
Contact silicide	0.35	Scaling—junction depth
CMP	0.35	MLM lithography, global dielectric
Cu metallization	0.18	R-C propagation delay
Dual damascene	0.18	Lithography—global planarization, Cu RIE process
Zero-overlay line via	0.18	Scaling
Low-κ dielectric	0.13	R-C propagation delay

Source: Adapted from Keith Buchanan, in *Proceedings of the 2002 GaAsMANTECH Conference,* San Diego, CA, April 8–11, 2002, table 1.

Notes: ARC, top anti-reflective coating; CMP, chemical-mechanical polishing or planarization; MLM, multi-level metal.

Many of the innovations noted in Table 9.1 were aimed at preventing metal diffusion, which was not much of a problem when trace width was over a micron wide but became very much an issue at submicron sizes and shallow junction depths.

The switch to copper interconnects started back in 1998, though still paired with silicon dioxide insulators. Finding the right dielectric material to pair with copper conductors to maximize electrical performance at the given feature-size node turned out to be more problematic than expected, though implementing copper metallization was not without its own complications.

For instance, copper can diffuse easily into silicon and into many other dielectric materials. Unlike aluminum, copper can also diffuse and ionize in the presence of electric fields. For these reasons, metallic diffusion barriers need to be added to the sidewalls of vias and lines prior to any copper plating or deposition. However, too thick a diffusion barrier layer in the trenches would elevate the interconnect's overall effective resistivity (barrier layer + copper metallization), so the layer must be kept thin to avoid that effect.

9.3.2   Dielectric materials

As previously noted, the typical insulator used for silicon CMOS was silicon dioxide for many years, with a dielectric constant, abbreviated as κ, of about 3.92. One distinct advantage for using silicon dioxide as an insulator is that the interface between the silicon and insulator can be created with minimal defects. Defects can trap electrical charges and adversely affect a nearby transistor's electrical behavior. Over those many years, process engineers acquired the experience to process silicon dioxide-on-silicon with very low defect densities.

Moving to low-κ took several baby steps. One of the first moves was using fluorine-doped silicon dioxide, with a κ of 3.5 to 3.7, at the 0.18-μm technology node. But switching from silicon dioxide to a truly low-κ (κ = 2.7 to 3) material has turned out to be much more challenging than anticipated, with regard to reliability and yield issues. Table 9.2 illustrates how the low-κ implementation kept getting pushed out over the years, as process engineers found ways to stick with trusty silicon dioxide, or some variation that did not require extensive equipment or process changes, and delay the leap to something unfamiliar for as long as possible. The first real attempts at widespread implementation did not occur until the 0.13-μm technology node around 2002.

Table 9.3 lists some of the challenges posed by integrating low-κ materials into wafer processing. Two of the factors listed, hardness/modulus and thermal stability, are of particular interest for assembly and packaging, as an indication of whether the devices can withstand the mechanical

Table 9.2 Delay in Low-κ Implementation

Year	1997 NTRS	1998 International Technology Roadmap for Semiconductors (2001–2002 ITRS) Update	1999–2000 ITRS	2001–2001 ITRS	2003–2004 ITRS
1999	3.0	>4.0	>4.0	—	—
2001	2.5	3.5	3.5	>3.5	—
2003	2.0	<2.5	>2.5	>3.5	>3.5
2005	1.5	2.0	>2.0	>3.0	
2007	—	>1.5	>1.5	>2.5	3.0
2009	—	—	—	>2.5	3.0
2011	—	—	—	>2.0	>2.5
2013	—	—	—	—	<2.5

Source: Adapted from Alexander E. Braun, "Low-κ Bursts Into the Mainstream... Incrementally," *Semiconductor International*, May 2005.

Table 9.3 Challenges Facing Low-κ Dielectric Integration

Low-κ Physical Property	Challenge
Hardness, modulus	CMP, packaging
Thermal conductivity	Reliability
Porosity	Etch/strip/barrier deposition
Chemical stability	Etch/strip compatibility
Thermal stability	BEOL process compatibility
Electrical stability	Reliability–leakage and breakdown
Adhesion	Reliability

Source: Adapted from Keith Buchanan, in *Proceedings of the 2002 GaAsMANTECH Conference*, San Diego, CA, April 8–11, 2002, table 1

Notes: CMP, chemical-mechanical polishing or planarization; BEOL, back-end-of-line.

and thermal stresses applied during back-end processing, without taking a severe final yield hit or affecting long-term reliability out in the field.

9.4 Dielectric constant requirements at each technology node

Figure 9.2 illustrates how the effective κ must be reduced as the wafer process technology goes from 65 nm to 45 nm. Reaching each goal has proven more and more difficult, though in the end, successful.

Assumptions

Cu D, B height = 35 nm
Hardmask height = N/A
Via height = 112 nm
Trench height = 126 nm
Minimum L/S = 70 nm

Assumptions

$K_{(Cu\ D,\ B)}$ = 4.5
$K_{(Hardmask)}$ = N/A
$K_{(Via)}$ = 2.9
$K_{(Trench)}$ = 2.9
K_{eff} = 3.2

Assumptions

Cu D, B height = 35 nm
Hardmask height = 40 nm
Via height = 112 nm
Trench height = 126 nm
Minimum L/S = 70 nm

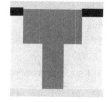

Assumptions

$K_{(Cu\ D,\ B)}$ = 4.5
$K_{(Hardmask)}$ = 4.1
$K_{(Via)}$ = 2.7
$K_{(Trench)}$ = 2.7
K_{eff} = 3.3

(a)

Assumptions

Cu D, B height = 35 nm
Hardmask height = 40 nm
Via height = 112 nm
Trench height = 126 nm
Minimum L/S = 70 nm

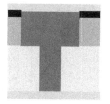

Assumptions

$K_{(Cu\ D,\ B)}$ = 4.5
$K_{(Hardmask)}$ = 4.1
$K_{(Via)}$ = 2.7
$K_{(Trench)}$ = 2.7
K_{eff} = 3.3

Assumptions

Cu D, B height = 30 nm
Hardmask height = N/A
Via height = 80 nm
Trench height = 90 nm
Minimum L/S = 50 nm

Assumptions

$K_{(Cu\ D,\ B)}$ = 4.0
$K_{(Hardmask)}$ = N/A
$K_{(Via)}$ = 2.7
$K_{(Trench)}$ = 2.7
K_{eff} = 3.0

Assumptions

Cu D, B height = 30 nm
Hardmask height = 35 nm
Via height = 80 nm
Trench height = 90 nm
Minimum L/S = 70 nm

Assumptions

$K_{(Cu\ D,\ B)}$ = 4.0
$K_{(Hardmask)}$ = 3.0
$K_{(Via)}$ = 2.5
$K_{(Trench)}$ = 2.5
$K_{(Middle-STP)}$ = 4.0
K_{eff} = 2.9

(b)

Assumptions

Cu D, B height = 35 nm
Hardmask height = 40 nm
Via height = 112 nm
Trench height = 126 nm
Minimum L/S = 70 nm

Assumptions

$K_{(Cu\ D,\ B)}$ = 4.0
$K_{(Hardmask)}$ = 3.0
$K_{(Via)}$ = 2.5
$K_{(Trench)}$ = 2.5
K_{eff} = 2.9

Figure 9.2 (a) Potential solutions at 65 nm—realistic case (2007, 2008). (b) Potential solutions at 45 nm—realistic case (2009, 2010, 2011).

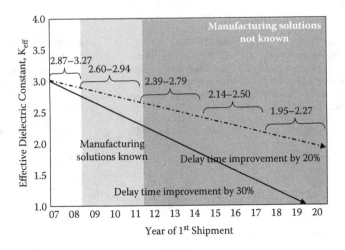

Figure 9.3 κ_{eff} Roadmap, ITRS 2007 revision.

9.5 Future interconnect and dielectric materials

In the long term, however, material innovations alone will not be enough to meet performance requirements while features sizes continue to scale down. But this will be a daunting task: according to ITRS, there are no manufacturing solutions known at this time once the technology node reaches 16 nm or 14 nm, at around 2012 (Figure 9.3).

9.5.1 Interconnects for <22 nm

Before long, copper performance will reach its limits and some novel interconnection solution must be found. As shown in Figure 9.4, effective

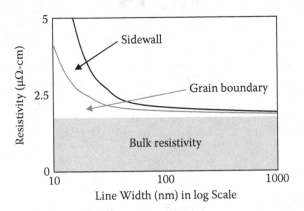

Figure 9.4 Copper (Cu) resistivity.

copper resistivity shoots up at line widths below 100 nm, as grain boundary and sidewall effects overtake the bulk resistivity. According to the 2007 Edition, the ITRS says no possible solution has been identified to deal with the phenomenon.

The ITRS expects at some point there will be the need to radically move away from the metal/dielectric materials systems that has served the semiconductor industry so well for decades. What the new architecture might look like is quite unknown at this time. Table 9.4 looks at possible paths interconnect technology might take.

One option noted might be placing more of the interconnect burden on package design and technology. The critical challenges to marry packaging with the integrated circuits more closely will be developing design tools that can model accurately this much more complex, intertwined system, the research and development of materials and technology to make such combinations possible, and the cost of ensuring reliability and performance meet requirements.

Table 9.4 Alternate Solutions for On-Wafer Interconnect Systems

Use different signaling methods	
	Signal design
	Signal coding techniques
Use innovative design and package options	
	Interconnect-centric design
	Package-intermediated interconnect
	Chip-package codesign
Use geometry	
	Three-dimensional (3D) structures
Use different physics	
	Optics—emitters, detectors, free space, waveguides
	Radio-frequency (RF)/microwave—transmitters, receivers, free space, waveguides
	Terahertz photonics
Radical solutions	
	Nanowires/nanotubes
	Molecules
	Spin
	Quantum wave functions

Source: Adapted from *International Technology Roadmap for Semiconductors, 2007 Edition,* Inteconnect chapter, table INTC5.

The other options listed are more wafer and chip oriented, but if they were implemented, they would have profound effects on back-end packaging and test, as the novel integrated circuit structures might not fit well with typical packaging technologies in use.

9.5.2   Dielectric materials for ≥22 nm

Currently, solutions are known and are either available or being developed for use in the next few years. The main difficulties lie in implementation and manufacturability. None of the solutions offer the simplicity of laying down bulk layers of silicon dioxide. Instead, multiple steps and layers of different dielectric materials—either a form of spin-on glass or deposited silica, a spin-on polymer, or even an intentional air gap—are laid down to achieve the desired effective κ constant. This is implied already in Figure 9.2, which shows the varied dielectric constants needed at via etch stop layer versus the hardmask (dielectric protection) layer versus the copper + diffusion barrier layer, and so forth.

9.5.3   Dielectric materials for <22 nm

The goal here is to bring the bulk material κ down below 2.5, which probably translates to an effective κ of 2.8 and below. There are big challenges to meet that goal, and they are often in opposition to each other during implementation.

For instance, to achieve a dielectric constant below 2.5, significant porosity must be incorporated in the insulator thin film, which tends to be a type of silica, usually with backbone modifications involving carbon. Obviously, a very porous interlayer dielectric will have several unintended effects. One is that the greater porosity requires an effort to manage the line and via sidewall roughness. However, that surface roughness has a detrimental effect on copper resistivity—at high frequencies, current flow remains on the surface, and a rough surface means the current must follow the contours, retarding rapid flow.

Another issue is that these porous silicas all exhibit coefficients of thermal expansion that end up placing the interconnect metallization in tension. This will likely result in "via popping." These failures may not show up until after final assembly and testing are completed. It turns out that having more carbon in the backbone raises thermal expansion coefficients, thus applying more stress to the metallic interconnects.

And, these low-κ materials all exhibit lower mechanical and yield strength levels compared to silicon dioxide, making them more susceptible to damage. Not surprisingly, adding porosity to said material only makes their mechanical behavior worse.

9.6   Future packaging options

Finding packaging solutions to meet these leading-edge chip requirements will demand a great deal of innovation and new ways of looking at
the problem from the industry.

9.6.1   Codesigning the chip with the package

One approach as to solutions required for increasing power and frequency
needs on the chip is to offload the requirements onto the parts outside
the chip, such as the package materials and design. For example, it may
be possible to move some interconnects off the chip and into the package, where thicker metallizations are allowable, and therefore, there are
higher performance levels. Signals could be routed through the package
and then back into the chip. This concept is related to the idea of simultaneously designing the chip with its package, where front-end and back-
end modeling and design tools would be united to allow characterization
of the entire system.

    Either goal poses some difficult challenges. Adding cost and complex-
ity to package design and materials will prove difficult to sell to customers.
Current design and modeling tools cannot handle this kind of multiscale,
multiphenomenon problem now, and it is unclear if-and-when such tools
could become available. It is unknown if cost-effective package materials
and designs can be created in either case. And last, it is not clear if much
or any of the chip's electrical performance requirements can be offloaded
effectively onto a package.

9.6.2   Three-dimensional (3D) integration

3D interconnection of active devices is an attractive solution to the problem of high-frequency signal propagation between chips. Stacking chips
physically and electrically may improve electrical performance while
reducing space and volume required in a system, when compared to a set
of stand-alone integrated circuits (ICs).

9.6.3   Through-silicon vias

Through-silicon via technology is considered to be a subset of wafer-level
packaging. Through-silicon vias allow for stacked monolithic chips to
communicate directly with one another, without the need for electrical
signals to travel along wire bonds or flip-chip bumps, through substrate
routing and onto another chip. As shown in Figure 9.5, such a configuration also reduces a package's footprint.

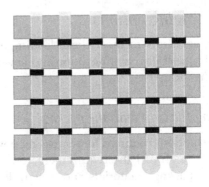

Figure 9.5 Stacked devices with through-silicon vias (TSVs).

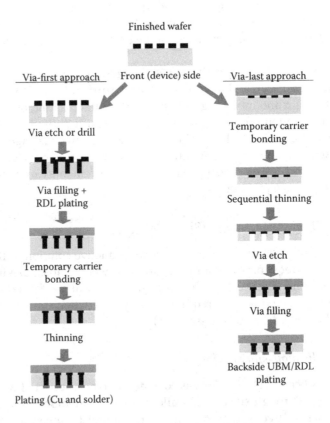

Figure 9.6 Through-silicon via process flow (postcomplementary metal oxide semiconductor [CMOS] processing) comparing via-first versus via-last. UBM, under bump metallization; RDL, redistribution layer.

Figure 9.6 illustrates the basic process flow for creating through-silicon vias after all front-end processing is completed. There is also research work being done to incorporate through-silicon via processing as part of wafer fabrication.

So far, through-silicon via has found use in certain niches in limited production—CMOS image sensors and Micro Electro-Mechanical System (MEMS). There are several technical challenges to widespread implementation of through-silicon via technology. These include high aspect ratio via etching, high speed—for production volumes—via filling methods, and equipment available to do precise wafer-level alignment and assembly in high-volume manufacturing. It may be years, or even a decade or more, before the design, logistical, manufacturing, and standards will be worked out fully and through-silicon via becomes a mainstream and high-volume packaging technology, joining wire-bonding, surface-mount, flip-chip, and area-array packaging.

Bibliography

A.E. Braun, "Low-κ Bursts Into the Mainstream…Incrementally," *Semiconductor International*, May 2005.

K. Buchanan, "The Evolution of Interconnect Technology for Silicon Integrated Circuitry," in *Proceedings of the 2002 GaAsMANTECH Conference*, San Diego, CA, April 8–11, 2002.

R. DeJule, "Infrastructure Still Inhibits 3-D ICs," *Semiconductor International*, May 1, 2009.

H.R. Huff and D.C. Gilmer, eds., *High Dielectric Constant Materials: VLSI MOSFET Applications*, pp. 483–485, Springer, New York, 2005.

International Technology Roadmap for Semiconductors, 2007 Edition, Assembly and Packaging chapter.

International Technology Roadmap for Semiconductors, 2007 Edition, Interconnect chapter.

International Technology Roadmap for Semiconductors, 2008 Update, Overview.

International Technology Roadmap for Semiconductors, 2009 Edition, Assembly and Packaging chapter.

International Technology Roadmap for Semiconductors, 2009 Edition, Interconnect chapter.

H. Johnson, "Surface Roughness," *EDN Magazine*, December 2001.

M. LaPedus, "TSV Chips: Not Ready for Prime Time," *EE Times*, June 10, 2010.

P. Marchal and M. Van Bavel, "Evaluating the Risks and Benefits of 3-D Technology," *Semiconductor International*, August 1, 2009.

B. Moyer, "Getting Around Limits by Getting High," *IC Design and Verification Journal*, December 2, 2008.

chapter ten

Light-emitting diodes

10.1 Objectives

- Give a brief description of a light-emitting diode (LED).
- Describe the unique packaging requirements of light-emitting diodes and their similarities to semiconductor assembly and packaging.
- Briefly discuss the reliability requirements of LED packages.

10.2 Introduction

A LED (light-emitting diode) is composed of a semiconductor chip filled with impurities to create what is known as a p-n junction, as shown in Figure 10.1. Current flows easily from the p-side (otherwise known as the anode) to the n-side (also known as the cathode) but not so in the opposite direction. The first LED products emerged in the 1960s, but in recent years they have found greater use due to many technological innovations, which will be described subsequently.

LEDs can generate more illumination per watt than incandescent bulbs, which is favorable for battery-powered or energy-saving devices. Also, LEDs can illuminate in colors without the need of additional filters. Table 10.1 shows the types and combinations of semiconductors needed to produce various colors of light, including white. Lastly, LEDs can focus their light independently, unlike incandescent and fluorescent sources, which need external equipment to collect and deflect light in a given direction.

High-brightness LEDs (light-emitting diodes) are emerging as an important development in illumination technology. LEDs offer a number of technical advantages:

- High efficiency converting electricity to light (>25% and growing)
- Fast response time (<1 microsecond)
- Physically small (500-µm square die)
- Long life (>30,000 hours)

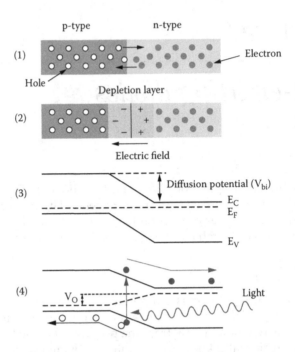

Figure 10.1 p-n Junction in (1)–(4). (From Wikimedia Commons.)

Table 10.2 shows a comparison of lifetime projections for white LED lights versus other light sources. So far it would appear LED lights can outlast most if not all other types of light sources.

The disadvantages are both economic—currently, high-brightness LEDs are initially more expensive than conventional sources such as incandescent and fluorescent bulbs, they have a relatively low lumen output, and the drive circuitry and power supplies required are more expensive than the older two technologies—and technological. For instance, so far LEDs remain relatively inefficient, as only 15% to 25% of the energy used gets generated into light, while the rest is dissipated as heat. And the issue of heat dissipation is a major difficulty in both assembling and operating LEDs. LEDs are adversely affected by elevated temperatures, and operating one in an overheated ambient environment for a prolonged period of time may result in overheating, which could lead to premature failure of the device.

However, in more and more cases, LED lights save money over the lifetime of the products, in spite of higher initial costs. The efficiency of converting power to light also continues to improve and is approaching that of fluorescent sources. The use of LED light sources in general lighting may accelerate in the middle of this decade, as governments at all levels around the world look to ban the inefficient—but very inexpensive—incandescent

Table 10.1 Semiconductors Types Used to Create Light-Emitting Diodes, and Their Respective Colors

Color	Potential Difference	Materials
Infrared	1.6 V	Aluminum gallium arsenide (AlGaAs)
Red	1.8 V–2.1 V	Aluminum gallium indium phosphide (AlGaInP), aluminum gallium arsenide (AlGaAs), gallium arsenide phosphide (GaAsP), gallium phosphide (GaP)
Orange	2.2 V	Aluminum gallium indium phosphide (AlGaInP), gallium arsenide phosphide (GaAsP)
Yellow	2.4 V	Aluminum gallium indium phosphide (AlGaInP), gallium arsenide phosphide (GaAsP), gallium phosphide (GaP)
Green	2.6 V	Aluminum gallium phosphide (AlGaP), aluminum gallium indium phosphide (AlGaInP), gallium nitride (GaN), indium gallium nitride (InGaN)
Blue	3 V–3.5 V	Gallium nitride (GaN), indium gallium nitride (InGaN), silicon carbide (SiC), sapphire (Al_2O_3), zinc selenide (ZnSe)
White	3 V–3.5 V	Gallium nitride (GaN) along with AlGaN quantum barrier present; gallium nitride (GaN) with indium gallium nitride (InGaN) active layer
Ultraviolet	3.5 V	Indium gallium nitride (InGaN), aluminum nitride (AlN), aluminum gallium nitride (AlGaN)

Source: Adapted from *Electronics Weekly*, December 5, 2008; and from Siang Ling Oon, "The Latest LED Technology Improvement in Thermal Characteristics and Reliability— Avago Technologies' Moonstone 3-in-1 RGB High Power LED," *Avago Technologies White Paper*, AV02-1752EN, March 17, 2010.

lightbulb. On the technology side, LED devices that can produce over 100 lumens per Watt are now more common, which is the level needed to make lights as bright as 40-Watt or 60-Watt incandescent light bulbs. The ongoing challenge is to continually reducing the total installed cost and, specific to this discussion, the cost of packaging the LEDs, while showing continuous improvement in illumination strength and other optical properties.

10.3   Unique characteristics of light-emitting diode (LED) packaging needs

LED chips are made from compound semiconductor materials that tend to be brittle and expensive as compared to silicon. Due to the difficulties in fabrication, individual LED devices are small—down to 250-μm squared

Table 10.2 White Light-Emitting Diode (LED) Lifetime Projections versus Other Types of Light Sources

Light Source	Typical Rated Lifetime Range (hours)[a]	Estimated Useful Life (L_{70})
Incandescent	750–2000	
Halogen incandescent	3000–4000	
Compact fluorescent (CFL)	8000–10,000	
Metal halide	7500–20,000	
Linear fluorescent	20,000–30,000	
High-power white LED		35,000–50,000

Note: L_{70} means a 70% lumen maintenance level.

Source: Adapted from U.S. Department of Energy, *Lifetime of White LEDs*, PNNL-SA-50957, June 2009.

[a] Varies by specific lamp type.

with a comparable height—fragile, and heat and environmentally sensitive. Assembly can be made more complicated by the shape of certain LED die, which may come in various polygon shapes rather than the typical rectangular slabs for semiconductors. Different geometries are used to minimize light absorption within the package. Unusual die shapes, however, require specific assembly processes and equipment. As already implied, current LED package assembly processes tend to be semicustom and often proprietary. One step that definitely differs from semiconductor packaging assembly is an addition of a yellow-tinted—typically based on yttrium aluminum garnet or YAG—phosphor powder on top of a blue or green-emitting diode, which then helps the LED give off the appearance of white light. Other phosphors may be used to aid LED diodes in giving off light at a given wavelength with greater stability or to tweak the range or spectrum of the light emitted, giving a "warmer" or "cooler" type of lighting. This particular process step using the phosphors tends to be a trade secret among the various LED manufacturers. Last, die attach material used for LED chips started with eutectic gold-tin (80%/20% Au-Sn), which is associated with ceramic or hermetic semiconductor packaging rather than now-mainstream plastic packages, though there is a move toward solders or even the silver-filled epoxy adhesives typically seen in plastic semiconductor packages.

LED packaging also requires the device/package structure to allow the maximum amount of light to exit in the pattern desired. Sometimes that may mean the LED chip sits within a highly reflective cavity, usually of a white color. The light extraction efficiency from small die is greater than from large die, because limited light is able to escape from the sides of large die. Light extraction is improved in flip-chip designs because no light is attenuated by semitransparent metal electrodes, absorption of light is dramatically reduced through the use of highly reflective metallization schemes, and no light is obscured by bond pads or wires.

Another major requirement of LED packaging is reliable heat dissipation. The packaging materials must be heat resistant and not subject to discoloration due to intense environmental changes, which would affect the light's reflectivity and intensity. Generally, the LED chip must be attached to a heat sink, heat spreader, or a heat-dissipating substrate, to prevent overheating issues. Heat-dissipating substrates may be ceramic—alumina, aluminum nitride (AlN), low-temperature co-fired ceramic (LTCC)—or perhaps an organic printed circuit board with a thick copper core. Sometimes, LED may be packaged in a modified transistor outline (TO) package or metal cans, but with transparent windows or lids. Figure 10.2 shows an example of a relatively simple LED package with through-hole pin leads.

Some products require multiple LED chips packaged together in one unit, often of different colors. The combination of red, green, and blue (RGB) LED in one package is increasingly popular, due to flexibility in

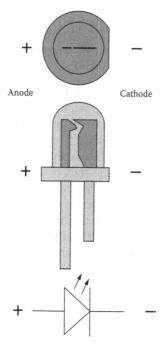

Figure 10.2 Example of a light-emitting diode (LED) package. (From Clker.com.)

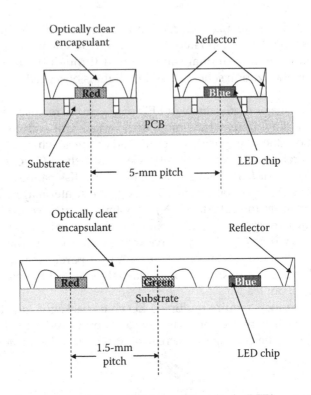

Figure 10.3 Comparison of discrete light-emitting diode (LED) component lighting matrix versus a red, green, blue (RGB) solution, in terms of pitch.

color selection and mixing along with illumination design. A typical RGB combination is a ratio of blue and green LED units to red ones of at least 2:1.

However, different colors mean different semiconductor compounds, which means different individual die. The colors and their corresponding materials are shown in Table 10.1. Thus, the assembly process must accommodate the resulting different die size, both wire bonding and flip-chip attach and sometimes different orientations. Equipment to "pick" these small, high aspect ratio die from many different wafers and place these die on a substrate within 25 μm of the desired location at high speed—at >1 die/second—is available typically only on a custom basis.

Still, combining multiple LED die into one package is attractive for a number of reasons. One is reducing the size of the lighting matrix needed to achieve good color quality, as illustrated in Figure 10.3.

10.4 Reliability requirements for LED packages

Table 10.3 shows a partial list of reliability requirements for LED devices. If compared to the tables and information provided in Chapter 5, the

Table 10.3 Examples of Reliability Requirements for Light-Emitting Diode (LED) Packages and Arrays

Reliability Test	Test Conditions	Test Length
High-temperature operating life (HTOL)	85°C case temperature at constant current	1000 hours
Room-temperature operating life	55°C case temperature	At constant current 1000 hours
Low-temperature operating life	–40°C case temperature at constant current	1000 hours
Temperature and humidity operating life	85°C case temperature and 85% relative humidity at constant current	500 hours
High-temperature storage life	120°C ambient	1000 hours
Low-temperature storage life	–40°C ambient	1000 hours
Temperature cycling	–40°C–120°C	1000 cycles
Power temperature cycling	–40°C–85°C at constant current	100 cycles

Source: Adapted from Bridgelux, Inc., *Application Note AN14: Reliability Data Sheet for Bridgelux LED Arrays*, April 7, 2009.

standards are similar, if not identical in many cases, and generally conform to many of the same corresponding JEDEC (formerly the Joint Electron Devices Engineering Council) standards.

Bibliography

Bridgelux, Inc., *Application Note AN14: Reliability Data Sheet for Bridgelux LED Arrays*, April 7, 2009.

S. Bush, "Lumileds Talks LED Packaging—Interview," *Electronics Weekly*, March 26, 2008.

P. Doe, "HB-LED Focus Turns to Cost Reduction and Next Generation Digital Intelligence," *SEMI Market Info*, March 1, 2011.

P. Doe, "$3 Billion Opportunity for Back-End Equipment and Materials Suppliers: HB-LED Packaging," *SEMI Market Info*, January 5, 2010.

D.D. Evans, Jr., "High Brightness Matrix LED Assembly Challenges and Solutions," *IMAPS 2007*, November 11–15, 2007.

International Technology Roadmap for Semiconductors, 2007 Edition, Assembly and Packaging chapter.

W.K. Jeung, S.H. Shin, S.Y. Hong, S.M. Choi, S. Yi, Y.B. Yoon, H.J. Kim, S.J. Lee, and K.Y. Park, "Silicon-Based, Multi-Chip LED Package," *Proceedings of the 57th Electronic Components and Technology Conference*, 722–727, Reno, NV, May 29–June 1, 2007.

L. Kim, W.J. Hwang, and M.W. Shin, "Thermal Resistance Analysis of High Power LEDs with Multi-chip Package," *Proceedings of the 56th Electronic Components and Technology Conference*, 1076–1081, San Diego, CA, May 30–June 2, 2006.

S. Kobilke, "Novel High Power Multichip LED Devices Give Customers Flexibility to Design-Their-Own Four-Chip LED Configuration," *LED Journal*, July/August 2007.

"LEDs: Your Electronics Weekly Guide," *Electronics Weekly*, December 5, 2008.

G. Mueller, R. Mueller-Mach, B. Grigory, R.S. West, P.S. Martin, T.-S. Lim, and S. Eberle, EU Patent No. WO2008104936: LED with Phosphor Tile and Overmolded Phosphor in Lens, September 4, 2008.

S.L. Oon, "The Latest LED Technology Improvement in Thermal Characteristics and Reliability—Avago Technologies' Moonstone 3-in-1 RGB High Power LED," *Avago Technologies White Paper*, AV02-1752EN, March 17, 2010.

P. Panaccione, T. Wang, X. Chen, S. Luo, and G.-Q. Lu, "Improved Heat Dissipation and Optical Performance of High-Power LED Packaging with Sintered Nanosilver Die-Attach Material," *Proceedings of the Sixth International Conference and Exhibition on Device Packaging*, March 9–10, 2010.

"Taiwan Makers Aim for More Efficient LED Lamp Packaging," *Taiwan Economic News*, June 5, 2009.

U.S. Department of Energy, *Lifetime of White LEDs*, PNNL-SA-50957, June 2009.

M. Wright, "Intematix Launches New Red and Green LED Phosphors," *LEDs Magazine*, November 11, 2010.

Glossary

3D bonding: A process joining two or more wafers or die surfaces together.

3D interconnection technology: Technologies that allow for the vertical stacking of electronic components that are connected by two-dimensional (2D) interconnect layers.

3D packaging: A general term referring to technologies where a substantial fraction of the die-to-die interconnections are not planar to the package substrate.

3D stacking: A 3D bonding process that also creates electrical interconnections between device layers.

3D System-in-package (3D-SIP): Three-dimensional (3D) integration using established package technologies, such as wire bonding or package-on-package stacking.

3D wafer-level packaging (3D-WLP): Three-dimensional (3D) integration using wafer-level package technologies, such as electrical interconnect redistribution.

Chemical mechanical polishing or planarization (CMP): Wafer cleaning method used on both metal and dielectric layers, often to prepare the surface for the next layer of material.

Embedded passives: Passive components that are incorporated into an integrated circuit (IC), added on top of an IC through the addition of a layer, embedded in a built-up polymer interconnect layer, or embedded in a package substrate.

Flip-chip ball grid array (FC-BGA): Similar to a plastic ball grid array (PBGA) where the die to substrate interconnection is made with the flip-chip process—that is, the die faces down with interconnection made through metal (solder) bumps on the die. Usually the space between the die and the substrate is filled with an underfill material.

Flip-chip land grid array (FC-LGA): Similar to FC-BGA, without the solder balls on the array of contact lands on the substrate.

Integrated passives: Integrated passives are arrays or networks of passive elements including resistors, capacitors, and inductors integrated on a single substrate to form a single passive component.

Mean time between failures (MTBF): The average time between failures in repairable or redundant systems.

Mean-time-to-failure (MTTF): The average time to failure for components or nonrepairable systems.

Plastic ball grid array (PBGA): A plastic package employing an array of solder balls for physical connection to the next level, which is usually a printed circuit ball. Usually the backside of the die is bonded to a laminate substrate, and the electrical connections are made on the active top side of the die through the wire bonding process. The top side of the package is encapsulated by a molding process.

Quad flat no-lead (QFN): A ceramic or plastic chip carrier with contact leads underneath the four sides of the package. Usually the backside of the die is bonded to the lead frame substrate, and the electrical connections are made to the die active surface through the wire bonding or flip-chip process.

Quad flat pack (QFP): A ceramic or plastic chip carrier with leads projecting down and away from all sides of a square package (i.e., gull-wing leads). Usually, the backside of the die is bonded to the lead frame substrate, and the electrical connections are made on the active top side of the die through a wire bonding process, and the whole package is encapsulated by a molding process.

Relative humidity (RH): The ratio of the amount of water vapor in the air to the maximum amount of water vapor that volume of air can hold at that temperature and pressure.

Second-level assembly: The attachment of a component to the next level of assembly packaging.

Substrate: The supporting material upon which or within which the elements of a semiconductor device are fabricated or attached.

System in package (SiP): System in package is characterized by any combination of more than one active electronic component of different functionality plus optionally passives and other devices like Micro Electro-Mechanical System (MEMS) or optical components assembled preferably into a single standard package that provides multiple functions associated with a system or subsystem.

Tape ball grid array (TBGA): Similar to PBGA where the substrate is made of a circuitized metal on a polymer tape. The interconnection to the die may be made by thermocompression bonding in a single step.

Through-silicon via (TSV): A technology under development to enable three-dimensional integrated circuit die stacking.

Through-silicon via (TSV) barrier layer: A barrier layer would be necessary to prevent diffusion of the TSV metal into the surrounding bulk silicon.

Under-bump metal (UBM): The metal layers located between the solder bump or column and the die.

"Via-first" through-silicon via (TSV) process: Fabricating the TSVs before the front end-of-line (FEOL) device fabrication steps begin.

"Via-last" through-silicon via (TSV) process: Fabricating the TSVs after FEOL processing is completed, and either after or during back end-of-line (BEOL) interconnect processing steps.

Wafer-level packaging (WLP): Wafer-level packaging (WLP) is a technology in which all the integrated circuit (IC) packaging and interconnection is performed on the wafer level prior to dicing. All elements of the package must be inside the boundary of the wafer. Chips mounted on a structured wafer—such as by face-to-face technologies—and packaged at wafer level before dicing are also considered as wafer-level packages.

Bibliography

C.A. Harper, *Electronic Packaging and Interconnection Handbook*, McGraw-Hill Professional, New York, 1991.

JEDEC, *JEP122E: Failure Mechanisms and Models for Semiconductor Devices*, March 2009.

JEDEC, *JEP150, Stress-Test-Driven Qualification of and Failure Mechanisms Associated with Assembled Solid State Surface-Mount Components*, May 2005.

JEDEC, *JEP156: Chip-Package Interaction—Understanding, Identification and Evaluation*, March 2009.

International Technology Roadmap for Semiconductors, 2007 Edition, Assembly and Packaging chapter.

International Technology Roadmap for Semiconductors, 2007 Edition, Interconnect chapter.

International Technology Roadmap for Semiconductors, 2009 Edition, Assembly and Packaging chapter.

Appendix A: Analytical tools

A.1   Introduction

Analytical tools are necessary to insure high-quality, high-reliability semiconductor packages for end market use, and are employed for quality assurance and reliability testing. And in the unlikely event package failures do happen, analytical tools will facilitate the failure analysis to determine the root cause.

A.2   Types of analytical tools

Basically, there are two categories to consider: destructive and nondestructive. Logically, it makes sense to employ nondestructive tests first, especially on live parts in a quality assurance situation or when the number of samples is limited to a few or one. Once nondestructive testing is exhausted, destructive tests could be turned to for further answers.

As a separate consideration, certain tests look at chemical or elemental analyses of a sample. Still others examine mechanical properties of the packaging materials prior to use, especially for behavior under stress or how the material might behave during manufacturing.

A.3   Nondestructive tools and tests

A.3.1   Introduction

These techniques preserve the integrity of the sample or samples for further examination, nondestructive or not. The following is a short list of typical nondestructive techniques for semiconductor package examination.

- Optical/visual inspection
- X-ray inspection
- Scanning acoustic microscopy

This appendix will mainly focus on scanning acoustic microscopy (SAM), a commonly used nondestructive analytical method, primarily for delamination and interface separation. Other methods touched on will include optical/visual and x-ray inspection methods.

A.3.2   Optical/visual inspection

Optical or visual inspection is used for incoming quality assurance (IQA) of packaging materials as well as the silicon chips. Parts are often examined visually during different stages of the assembly process as well as part of the final inspection procedure. This is also often the first step in failure analysis, examining the package and its leads visually, whether with the naked eye or a magnifying loupe or with an optical microscope. The failed part may still be attached to the printed circuit board or be alone when examined for external flaws. If failure proves to have external evidence, like the solder joint separation at the leads or a crack in the package body, that will point to the next step and kinds of analytical tools needed to probe for the root cause.

A.3.3   X-ray inspection

X-rays are a type of electromagnetic energy, along the spectrum with visible light but with shorter wavelengths (0.1 Å to 100 Å) and of higher energy. The properties of x-ray radiation allow the wavelengths to penetrate solid materials. Materials of lower density allow more x-rays to pass through than higher-density materials. Therefore, denser materials such as metal interconnect traces and bonding wires show up as darker shadows than do organic package materials in x-ray images.

The shorter the wavelength, the more readily x-rays will pass through solid materials. Control of the output wavelength is by setting the potential, or kV, of the x-ray source. The higher the kV, the shorter the wavelength, and the greater is its penetrating capability. Most x-ray inspection

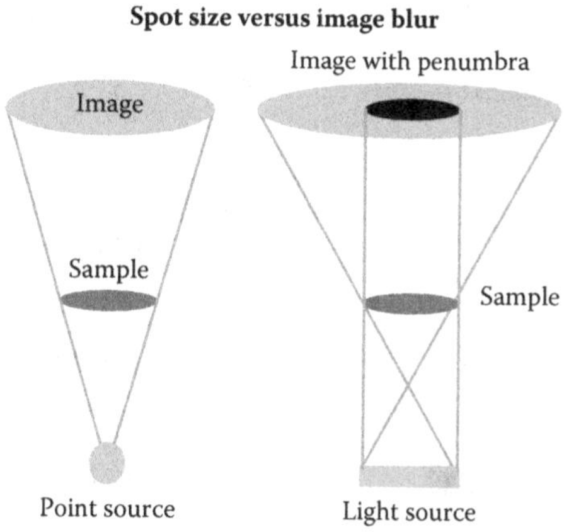

Spot size versus image blur

Figure A.1 X-ray resolution capabilities.

systems aimed at semiconductor packages and printed circuit boards (PCBs) have sources capable of operating at 75 kV to 160 kV.

Obtaining a good x-ray image requires a balance of optimizing wavelength (or kV) to provide desired contrast between dense and less-dense materials, and current (mA) to control the image brightness.

Another important parameter for optimal imaging is the spot size of the x-ray source. The spot is the area on the source where the x-rays are generated. The smaller the spot, the better the resolution capability will be, due to a phenomenon called penumbral blur (Figure A.1). However, the spot size is limited by the power required, because heat is also generated at the x-ray source. If too much energy is focused on a small spot, the heat will damage the source. Most systems will vary the spot size with x-ray power (the product of kV × mA) to protect and maximize the useful life of the source.

X-ray systems allow for a high magnification of the inspected sample. Geometric magnification is defined as the ratio of the distance between the source and detector to the distance between the source and the sample (Figure A.2). Obtaining the best resolution under high magnification requires a high level of geometric magnification. This can be achieved using a large distance between the source and detector, but that requires a large cabinet and is often not practical. Therefore, the most effective way to achieve high geometric magnification is to place the sample as close to the source as possible.

The trade-off for high magnification is field-of-view, though; the greater the magnification, the smaller the field-of-view. Different areas of concern require different levels of magnification. Looking at bonding wires may exceed 200X to 400X, for instance.

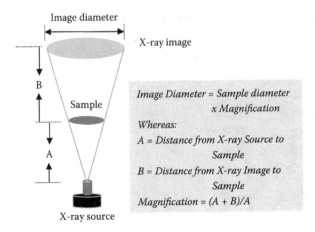

Figure A.2 Geometric magnification equation for x-ray imaging.

Given the behavior of x-rays, this analytical method is often used to look for wire sweep or interconnect misalignment. However, it is more difficult to detect an air gap or delamination with x-rays, because x-rays do not change properties that much going through an air gap. Makers of x-ray inspection systems have developed tools to overcome some of the tools' shortcomings, however. Image processing can overlay color or produce three-dimensional (3D) images. Air gap or void analysis is possible by distinguishing the levels of grayscale in the image through software analysis.

A.3.4   Scanning acoustic microscopy

As already noted, scanning acoustic microscopy (SAM) is often used to check package integrity, whether for quality assurance or in failure analysis. Again, it is a nondestructive imaging technique, using reflected or transmitted ultrasound signals to make visible images of internal features, including detecting defects like air pockets and areas of delamination within the package.

The sound waves are then absorbed, scattered, or reflected as they are aimed at a sample. The waves' response determines how the visible image is created. The only real requirement in sample preparation is that the sample can be submersed in water or another fluid, as air is a very poor transmitter of high-frequency acoustic waves, from 5 MHz to over 400 Mhz. By contrast, the human ear rarely detects sound above 20 KHz.

Given that high-frequency sound waves cannot travel across air gaps as narrow as below 1 micron makes SAM a perfect means to look for areas of poor adhesion, as between the substrate and molding compound. The sound wave gets reflected nearly 100% when hitting an air pocket, while those passing through a solid sandwich compound, and the PCB would always see some level of transmission. That is why SAM has been so effective in detecting defects (cracks, delamination, voids, etc.) in plastic semiconductor packages, because the gaps are typically filled with air.

As a rule of thumb, high frequency means short focus, and low frequency means long focus. The higher the frequency used, the higher the possible resolution, but at the sacrifice of depth of field. The lower-frequency transducers are more than adequate for older, leaded plastic packages—plastic dual in-line packages (PDIPs), plastic quad flat packs (PQFPs), and small outline integrated circuits (SOICs). Higher frequencies are needed to resolve the finer features inside plastic ball grid arrays (PBGAs). The highest frequencies are required to resolve the submicron features in chip scale packaging (CSP) and flip-chip packaging.

To image a sample, typically the ultrasonic transducer raster-scans the sample, with thousands of acoustic pulses entering the sample per second. What is called an A-scan (Figure A.3) is rendered as a waveform display of what the transducer sees at that particular point in the sample (Figure A.4). Each A-scan pulse is described by its polarity—whether positive or negative—amplitude, position in time (which relates to depth), and frequency content. In standard imaging modes, the polarity, amplitude, and time data are used.

The amount reflected and transmitted depends on the acoustic impedance (Z) of the material and in contrast to the material it is coupled to. The greater the difference in Z, the more the ultrasound is reflected, which is illustrated in Figure A.5. As can be seen in Table A.1, air is considered to have zero acoustic impedance and causes complete reflection of the sound waves.

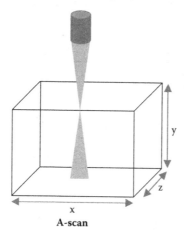

A-scan

Figure A.3 A-scan in scanning acoustic microscopy.

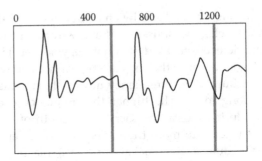

Figure A.4 Waveform display corresponding to an A-scan in scanning acoustic microscopy.

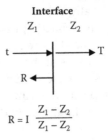

Figure A.5 Effect of acoustic impedance on coupled materials.

Table A.1 Acoustic Impedance of Materials Commonly Used in Semiconductor Packaging

Materials	Acoustic Impedance (Z)
Tungsten (W)	104
Alumina (Al_2O_3)	21 to 45
Copper (Cu)	42
Beryllia (BeO)	32
Silicon (Si)	20
Aluminum (Al)	17
Glass	15
Plastic	2.0 to 3.5
Water (H_2O)	1.5
Air/vacuum	0

Source: Adapted from Sonix, ABC's of Ultrasonics, presentation.

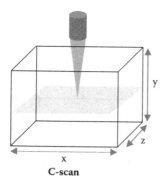

C-scan

Figure A.6 C-scan in scanning acoustic microscopy.

Not all the acoustic signals reflected back are used to create a visible planar acoustic image. Generally, such an image is generated of a particular interface or depth in the sample, in what is usually called a *C-Scan* in the *x-y* plane of the sample (Figure A.6). To create this image, a "time window" is set up to only allow return echoes from a particular depth in the sample, which is also known as "gating" of the return echoes (shown as the thick, dark gray lines in Figure A.4). This is possible because reflections from different depths will return to the transducer at different times, so it is possible to filter out those that are not of interest.

C-Scan focuses on a particular *x-y* plane at particular *z*-depth in a reflection mode (the transducer sends out and also picks up the reflected pulse). The depth of field is shallow. In other words, if the C-scan is focused on the die attach layer, it is unlikely to pick up any information about pad backside delamination.

As for the minimum size of defects detectable, it depends according to the suppliers of SAM machines. With a 15-MHz transducer, air gaps as narrow as 0.13 µm can be detected, and as the frequency goes up, the gap can be narrower. The detectable minimum size of a defect depends on the beam diameter (BD). BD is estimated by the following formula in Equation (A.1):

$$BD = 1.028 \, (FL \times C) / (f \times D) \qquad (A.1)$$

where
FL is the focal length; C is the velocity of material; F is the frequency of transducer, and D is the diameter of piezoelectric crystal.

C-Scan is a particularly effective way to find tiny defects but can also be time-consuming, and its effectiveness depends on the skill of the operator. It is most useful as a failure analysis (FA) tool. For quality assurance, a different way to use SAM is recommended. This method, called

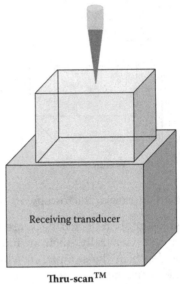

Figure A.7 THRU-Scan™ in scanning acoustic microscopy.

THRU-Scan™ (trademarked by Sonoscan), uses ultrasound in a trans-mission mode, as shown in Figure A.7. One transducer sends ultrasound waves through a specimen, while a receiving transducer picks up signals on the other side. Because voids and other air space defects cannot trans-mit ultrasound, the loss of signal at the receiver indicates the presence of defects. The same goes for interlayer delamination, such as the separation between a substrate's solder mask and the underfill material used in a flip-chip package.

C-Scan and Thru-Scan are just two examples of imaging modes avail-able in SAM. SAM can be used to image various layers and cross sections in a given sample. In all, SAM has proven to be an effective tool for exam-ining electronic packages nondestructively, whether for quality control, reliability, or failure analysis purposes.

Bibliography

W.E. Bernier, "Flip Chip PBGA Assembly: Quality and Reliability Challenges," presented at IMAPS, October 2, 2008.
C. Cohn and C.A. Harper, *Failure-Free Integrated Circuit Packages: Systematic Elimination of Failures Through Reliability Engineering, Failure Analysis, and Material Improvements*, McGraw-Hill Professional, New York, 363 pp, 2004.
C.A. Harper, *Electronic Packaging and Interconnection Handbook*, McGraw-Hill Professional, New York, 1991.

Renesas Technology, *Semiconductor Reliability Handbook*, REJ27L0001-0101, Rev. 1.01, 177, November 28, 2008.

Sonoscan, Inc., www.sonoscan.com/index.html

Wikipedia, entry on acoustic microscopy: http://en.wikipedia.org/wiki/Acoustic_microscopy

Sonix, http://www.sonix.com/index.php3

Sonix, Minimum Detectable Defect Size, presentation, 2002.

Sonix, ABC's of Ultrasonics, presentation.

YXLON International, http://www.yxlon.com/semiconductor_packaging

D. Naugler, "X-ray Inspection Technology: Finding Hidden Defects," *Advanced Packaging*, September 2006.

Appendix B: Destructive tools and tests

B.1   Introduction

These techniques, on the other hand, essentially alter or destroy the sample, irreversibly. Following are some of the commonly employed methods with respect to semiconductor packaging:

- Decapsulation techniques
- Dye penetration for leak detection
- Cross-sectioning and polishing
- Scanning electron microscopy (SEM)
- Transmission electron microscopy (TEM)
- Chemical and elemental testing
- Other analytical techniques

A few other test methods were discussed in previous chapters. First from Chapter 6, Section 6.1.5 was flexure testing to determine flexural modulus. Next from Chapter 6, Section 6.2.4, is a brief description of differential scanning calorimetry (DSC). Finally from Chapter 7, Section 7.2.7, three examples of destructive tests for bonding wires are discussed:

- Bond etching
- Bond pull
- Ball shear tests

B.2   Decapsulation

In the case of the plastic semiconductor packages, decapsulation generally means removing the molding compound from the top of the package, by mechanical grinding or through plasma or acid etching or a combination of two or more methods, to expose the chip surface and package structures, like the bonding wires and their corresponding bonds, in order to facilitate failure analysis by other techniques, such as SEM or bond pulls, depending on the suspected failure mode.

In all cases, care must be taken during the decapsulation process to preserve the internal features. Thus, the mechanical grinding of the molding compound must not get too close to the chip or substrate surface, for instance. The chemical or plasma etching must not be too aggressive lest the evidence be removed.

B.3   Dye penetration

Placing semiconductor package samples in a fluorescent dye bath under pressure is a long-established method to detect package cracking. Package cracks allow moisture to enter inside the package and cause all sorts of reliability issues, like corrosion. The dye will be forced into any cracks and will illuminate when the samples are placed under an ultraviolet (UV) light.

B.4   Cross-sectioning and polishing

Cross-sectioning a package is pretty much self-explanatory: the sample is cut through its y-z axis at a specific location to examine its interior construction and, hopefully, find the defect or defects suspected to have caused failure. Usually some amount of mechanical polishing is performed on the sample to better bring out the features, which may be further augmented with chemical polishing and etching.

B.5   Scanning electron microscopy (SEM)

Scanning electron microscopy creates a magnified image by using electrons instead of light waves. With electrons, very detailed three-dimensional images are possible at much higher magnifications than are possible with visible light. SEM is not a destructive technique per se, but it does require the sample to be irreversibly exposed and altered for visual analysis.

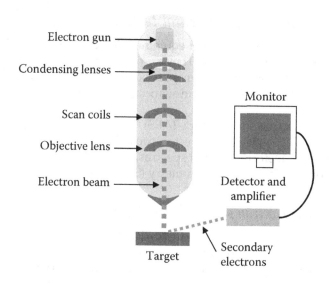

Figure B.1 Scanning acoustic microscope schematic.

For one thing, the sample must be prepared so that it withstands the vacuum inside the microscope. Another factor in sample preparation is that it must be able to conduct electricity in order to be illuminated by electrons. Nonconductive samples may need gold sputtered upon them to render them conductive. Finally, the sample must be able to withstand electron beam bombardment.

Once the sample is placed inside the vacuum chamber of the SEM, an electron gun emits a beam of high-energy electrons down through a series of magnetic lenses designed to focus the electrons to a given small area. As the electron beam reaches the sample surface, secondary electrons are released from the sample, which are then detected and their signals amplified. The final image is built from the number of electrons emitted from each point on the sample. This is illustrated in Figure B.1.

B.6 Transmission electron microscopy (TEM)

Transmission electron microscopy, and the related technique scanning transmission electron microscopy, both utilize an electron beam to image a sample, much like a SEM but with much higher spatial resolution, on the order of angstroms. Greater material characterization information can be obtained through TEM, such as crystallographic orientation, but sample preparation can be time-consuming, and samples may not be able to withstand the strong electron beam.

B.7   Chemical and elemental tests

These techniques and associated equipment look at the chemical and elemental composition of a sample. They are often employed to look for surface contamination, as they cannot penetrate very far beneath the exposed layer. A partial list of techniques used in semiconductor packaging is as follows:

- Auger electron spectroscopy (AES)
- Energy-dispersive x-ray spectroscopy (EDS or EDX)
- Fourier transform infrared spectroscopy (FTIR)
- Secondary ion mass spectrometry (SIMS)

B.7.1   Auger electron spectroscopy (AES)

Auger electron spectroscopy (AES), is used to identify chemical elements on the surface of materials. Like SEM, analysis is performed under a vacuum, and some level of electrical conductivity is necessary for analysis. Of course, coating the sample for conductivity would negate the possibility for chemical analysis.

AES is considered an excellent technique for analyzing the chemical composition of solid surfaces due to its high sensitivity for chemical analysis in the 5- to 20-Å region near the sample surface, relatively rapid and high-resolution data detection, and the ability to detect all elements above helium.

As shown in Figure B.2 using titanium as the sample material, the basic Auger process starts with removal of an inner-shell atomic electron to form a vacancy. In the case of AES, the vacancy is produced by bombarding the sample with an electron beam. The inner shell vacancy is filled

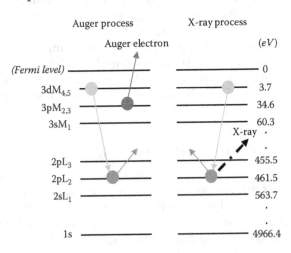

Figure B.2 Comparison of Auger electron spectroscopy versus x-ray.

by a *second* atomic electron from a higher shell. Energy must be simultaneously released. A *third* electron, the Auger electron, escapes carrying the excess energy in a radiationless process. The process of an excited ion decaying into a doubly charged ion by ejection of an electron is called the Auger process. Alternatively, an x-ray photon removes the energy.

As previously stated, the procedure for AES is similar to that for SEM (which was illustrated in Figure B.1) and is as follows: the sample is irradiated with electrons from an electron gun, and the emitted secondary electrons are analyzed by an electron spectrometer.

B.7.2 Energy-dispersive X-ray spectroscopy (EDS or EDX)

Energy-dispersive x-ray spectroscopy is an analytical technique often coupled with imaging tools like SEM. The impact from the electron beam on the sample also generates x-rays that are characteristic of the elements present, as hinted in Figure B.3. Naturally, the samples should not be coated for conductivity for this technique to work.

B.7.3 Fourier transform infrared spectroscopy (FTIR)

Fourier transform infrared spectroscopy (FTIR) allows for chemical and elemental detection in a sample. Infrared light is passed through a material, and the resulting spectrum after the light is selectively absorbed and transmitted by the sample indicates the composition, distribution, and even the quantities of materials present.

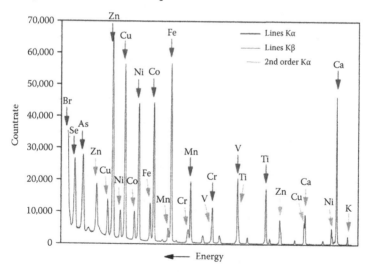

Figure B.3 Example of energy-dispersive x-ray spectroscopy. (From Wikimedia Commons.)

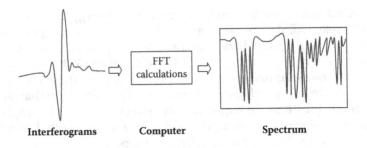

Interferograms   Computer   Spectrum

Figure B.4 Fourier transform of an interferogram into a spectrum.

FTIR has advantages over the older techniques using dispersive technology. FTIR can read all the frequencies of energy transmitted simultaneously using an interferometer. The resulting signals, called interferograms, cannot be interpreted directly and must undergo Fourier transformation to provide the desired spectral information for analysis. This is illustrated in Figure B.4.

The sample analysis process is next shown in Figure B.5. An infrared beam passes through the interferometer on its way to the sample. A detector absorbs the energy signals after passing through the sample. The signals are then processed and transformed by the computer into a spectrum, which can then be analyzed.

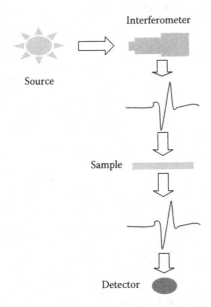

Figure B.5 Sample analysis with Fourier transform infrared spectroscopy.

B.7.4   Secondary ion mass spectrometry (SIMS)

Secondary ion mass spectrometry (SIMS) is a technique to detect minute amounts of dopants and impurities at or below the sample surface, to a maximum of tens of microns. SIMS works by sputtering the sample surface with a beam of primary ions. The secondary ions given off during the sputtering are analyzed with a mass spectrometer.

All elements and their isotopes can be detected, at ppm levels or less. However, the data are only at the elemental level and do not give any indication about chemical compounds or bonding. Last, samples must be able to be in a vacuum environment for evaluation.

B.8   Other analytical techniques

The test methods summarized here look at determining material properties, whether they are physical, chemical, thermal, or mechanical. As mentioned at the beginning of this appendix, several of the analysis techniques discussed here were already touched upon in the preceding chapters. These include the following:

- Bonding wire pull
- Ball bond shear
- Differential scanning calorimetry (DSC)
- Flexure testing

B.8.1   Bonding wire pull

The bond-pull test is the primary method used for optimizing the bonding window and monitoring the bond quality. As already shown in Figure 7.10, Figure B.6 illustrates the testing setup and methodology. An upward force is exerted on the wire length, through the use of a pull hook, and force value required to break the wire, either in the bulk or at the bonding points, is recorded—typically in grams-force. The pull hook is usually positioned at the highest point in the wire loop. It should be understood that the pull test is influenced by package configuration and wire length. Test results include bond strength and failure mode. Breakage at either bond location—first (ball) bond or second (stitch) bond—is considered to be a sign of serious issues that will require further investigation.

B.8.2   Ball bond shear

Shear tests have long been used to determine the quality of a wire ball bonded to the bond pad. The amount of force required to push a ball bond off the pad is expected to correlate to the adhesion strength of the weld

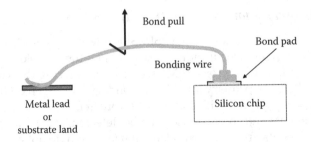

Figure B.6 Test setup for bond pull.

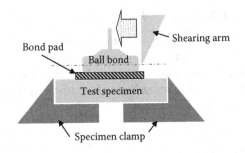

Figure B.7 Cross section of ball bond shear test setup.

between wire and pad. An example of wire ball bond shear test setup is shown in Figure B.7 (taken from Figure 7.11). Care must be taken to properly position the shear tool against the ball alone, in order to produce valid results.

B.8.3 *Differential scanning calorimetry (DSC)*

The purpose of a differential scanning calorimetry (DSC) test is to determine the peak exothermic reaction temperature of a given polymer material, such as a die attach adhesive or a molding compound. The purpose of the DSC test is to determine the minimum cure temperature for a thermosetting polymer system—the point where cross-linking begins—along with other changes to the chemical or physical state.

An example of a DSC curve is repeated from Figure 6.14 as Figure B.8.

B.8.4 *Flexural testing*

As stated previously, the flexure test is a measure of the behavior of materials under simple beam loading or bending. Maximum stress and strain over increasing loads are measured and plotted in a stress–strain

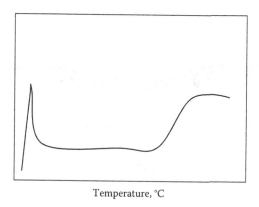

Temperature, °C

Figure B.8 Example of a differential scanning calorimetry curve (not to scale).

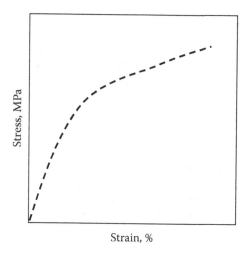

Strain, %

Figure B.9 Stress–strain curve from flexure testing.

diagram, as shown again in Figure B.9 (taken from Figure 6.5). Flexural strength is defined as the maximum stress endured in the outermost fiber of the material, which is calculated at the test specimen surface on the convex or tension side. Flexural modulus comes from calculating the slope of the stress versus the deflection curve.

Flexure testing is done on molding compounds to determine a composition's mechanical properties. Test methods include three-point flex and four-point flex. The three-point flex example is shown in Figure B.10 (taken from Figure 6.6), as this test method is most commonly used for polymers. In a three-point test, the area of uniform stress is rather small and concentrated under the center loading point.

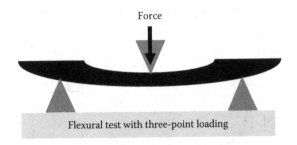

Figure B.10 Flexural stress test using three-point loading.

Bibliography

W.E. Bernier, "Flip Chip PBGA Assembly: Quality and Reliability Challenges," presented at *IMAPS*, October 2, 2008.

A.R. Chourasia and D.R. Chopra, Chapter 42: Auger Electron Spectroscopy in *Handbook of Instrumental Techniques for Analytical Chemistry*, 791–808, Prentice Hall, Upper Saddle River, NJ, 1997.

C. Cohn and C.A. Harper, *Failure-Free Integrated Circuit Packages: Systematic Elimination of Failures Through Reliability Engineering, Failure Analysis, and Material Improvements*, McGraw-Hill Professional, New York, 363 pp, 2004.

R.H. Estes, "A Practical Approach to Die Attach Adhesive Selection," *Hybrid Circuit Technology*, June 1991.

Evans Analytical Group, Technique Note, TN 101: EAGLABS[SM] Auger Electron Spectroscopy (AES) Services, Version 2.0, October 14, 2008.

Evans Analytical Group, Auger Electron Spectroscopy, www.eaglabs.com/techniques/analytical_techniques/aes.php

Evans Analytical Group, Energy Dispersive X-ray Spectroscopy, www.eaglabs.com/techniques/analytical_techniques/eds.php

Evans Analytical Group, Secondary Ion Mass Spectrometry, www.eaglabs.com/techniques/analytical_techniques/sims.php

Evans Analytical Group, Transmission Electron Microscopy and Scanning Transmission Electron Microscopy, www.eaglabs.com/techniques/analytical_techniques/tem_stem.php

C.A. Harper, *Electronic Packaging and Interconnection Handbook*, McGraw-Hill Professional, New York, 1991.

T. Hazeldine and K. Duong, "Microsurgery for Microchips—New Techniques for Sample Preparation," *Materials World*, vol. 11, no. 11, pp. 10–12, November 2003.

Instron, Flexure Test, www.instron.us/wa/applications/test_types/flexure/default.aspx?ref=http://www.google.com/search

K. Janssens, *Chapter 4: X-ray based methods of analysis in Comprehensive Analytical Chemistry XLII*, pp. 129–226, Elsevier B.V., New York, 2004.

B. Lee, "Electroless CoWP Boosts Copper Reliability, Device Performance," *Semiconductor International*, July 1, 2004.

Museum of Science, Boston, How the SEM Works, www.mos.org/sln/SEM/seminfo.html

D.E. Newbury, "The Revolution in Energy Dispersive X-Ray Spectrometry: Spectrum Imaging at Output Count Rates Above 1 MHz with the Silicon Drift Detector on a Scanning Electron Microscope," *Spectroscopy*, August 24, 2009.

Renesas Technology, *Semiconductor Reliability Handbook*, REJ27L0001-0101, Rev. 1.01, p 177, November 28, 2008.

G.E. Servais and S.D. Brandenburg, "Wire Bonding—A Closer Look," presented at ISTFA'91, Los Angeles, CA, November 11–15, 1991.

ThermoNicolet, *Introduction to Fourier Transform Infrared Spectroscopy*, P/N 169-707500, February 2001.

YXLON International, www.yxlon.com/semiconductor_packaging

Index

Printed in the United States
by Baker & Taylor Publisher Services